AVIATION MANAGEMENT
Global and National Perspectives

AVIATION MANAGEMENT

GLOBAL AND NATIONAL PERSPECTIVES

RATANDEEP SINGH

KANISHKA PUBLISHERS, DISTRIBUTORS
NEW DELHI-110 002

KANISHKA PUBLISHERS, DISTRIBUTORS
4697/5-21A, Ansari Road, Daryaganj
New Delhi -110 002
Phones : 2327 0497, 232 8 8285
Fax : 011-2328 8285
E-mail : kanishka_publishing@yahoo.co.in
kanishkabooks@gmail.com

Aviation Management:
Global and National Perspectives

First Published-2008

Edition-2024

ISBN : 978-81-8457-033-5

PRINTED IN INDIA

Published by Madan Sachdeva for Kanishka Publishers, Distributors, 4697/5-21A, Ansari Road, Daryaganj, New Delhi-110002, Typeset by Sunshine Graphics, Delhi, and Printed at G.S. Printers, Delhi.

Preface

This book entitled "Aviation Management: Global and National Perspectives" provides readers with an integrated picture of the subject area of aviation management so that a holistic understanding of this field of study could be created. A brief introductory of the subject area of aviation management is provided with focus on aviation. Civil aviation, airport, air traffic control, flight data recorder, airline, etc. Aviation sector in India has received special mention, so has been the case study on United Airlines Flight 232. The organizations related to aviation management have been dealt in detail with special mention of International Civil Aviation Organization, Aeropol Aviation Services Corporation, Aviation Management Consulting Group, AOPA, International Association of Airport Executives, etc. Various regulations and policy initiatives related to aviation management has been dealt in detail. Focus also lies on Federal Aviation Interactive Reporting System, Aircraft Regulations and Guidance, Convention of International Civil Aviation, Inter-Agency Committee for Aviation Policy, etc. Details are also documented regarding Active Level of Services Reviews, Aircraft Engineers International Affiliation and AVSEC Rules and Regulations. Case Study of BCAS India is briefly discussed. An overview of air travel in India is dealt.

The next chapter in this book deals with the important theme of commercial aviation management programmes, education and training. Case Studies are provided herein regarding subject areas of aviation management, flight operations and commercial aviation management. Relevant handbook programmes and specialization studies are also included. Special international case of Global Air Training is done in detain focusing mainly on vital issues such as crew resource management, team resource management, safety and security training. The study pattern followed by BCAS Training in India is also dealt. A detailed chapter is given on the frequently asked questions discussing the subject areas of aviation management, aviation technology, flight programme, aeronautical, flight engineering and airline water supplies. The next chapter deals with various air safety standards, guidelines and announcements with additional inputs on related aviation policy issues. A concluding chapter deals with contemporary status of global aviation industry. The Indian aviation industry and its future prospect is also discussed in brief. A detailed appendix is also given with details on prefixes and listings on Aircraft Registration, Airline Call-Signs and International Air Transport Association. An annotated bibliography is provided so that future referencing could be done effectively. In sum, this book will serve the purpose of a textbook on the subject.

Ratandeep Singh

Contents

1

Introduction to Aviation Management

INTRODUCTION TO AVIATION

Aviation refers to flying using aircraft, machines designed by humans for atmospheric flight. More generally, the term also describes the activities, industries, and regulatory bodies associated with aircraft.

History

There are records of early short-distance flights from the 9th to 11th centuries, such as the parachute flight of Armen Firman, the glider flight of, and possibly earlier human-carrying kites from China.

The modern age of aviation began with the first untethered human lighter-than-air flight on November 21, 1783, in a hot air balloon designed by the Montgolfier brothers, and balloon flight became increasingly common over longer and longer distances throughout the 19th century, continuing to the present.

The practicality of balloons was limited by the fact that they could only travel downwind. It was immediately recognized that a steerable, or dirigible, balloon was required. Although several airships, as steerable balloons came to be called, were built during the 1800s, the first aircraft to make routine flights were made by the Brazilian aviation pioneer Alberto Santos-Dumont. Santos-Dumont effectively combined an elongated balloon with an internal combustion engine. On October 19, 1901 he became world famous when he flew his airship "Number 6" over Paris to win the Deutsch de la Meurthe prize. Santos-Dumont's success with airships proved that controlled and sustained flight was possible.

On 1903, the Wright brothers flew the first successful powered, heavier-than-air flight, though their aircraft was impractical to fly for more than a short distance because of control problems. The widespread adoption of ailerons made aircraft much easier to manage, and only a decade later, at the start of World War I, heavier-than-air powered aircraft had become practical for reconnaissance, artillery spotting, and even attacks against ground positions.

Aircraft began to transport people and cargo as designs grew larger and more reliable. In contrast to small non-rigid blimps, giant rigid airships became the first aircraft to transport passengers and cargo over great distances. The best known aircraft of this type were manufactured by the German Zeppelin company.

The most successful Zeppelin was the Graf Zeppelin. It flew over one million miles, including an around the world flight in August of 1929. However, the dominance of the Zeppelins over the airplanes of the that period, which had a range of only a few hundred miles, was diminishing as airplane design advanced. The "Golden Age" of the airships ended on 6 June, 1937 when the Hindenburg caught fire killing 36 people. Although there have been periodic initiatives to revive their use, airships have seen only niche application since that time.

Great progress was made in airplane design during the 1920s and 1930s. One of the most successful designs of this period was the Douglas DC-3 which became the first airliner that was profitable carrying passengers exclusively, starting the modern era of passenger airline service. By the beginning of World War II, many towns and cities had built airports, and there were numerous qualified pilots available. The war brought many innovations to aviation, including the first jet aircraft and the first liquid-fueled rockets.

After World War II, especially in North America, there was a boom in general aviation, both private and commercial, as thousands of pilots were released from military service and many inexpensive war-surplus transport and training aircraft became available. Manufacturers such as Cessna, Piper, and Beechcraft expanded production to provide light aircraft for the new middle class market.

By the 1950s, the development of civil jets grew, beginning with the de Havilland Comet, though the first widely-used passenger jet was the Boeing 707. At the same time, turboprop propulsion began to appear for smaller commuter planes, making it possible to serve small-volume routes in a much wider range of weather conditions.

Yuri Gagarin was the first human to travel to space on April,

1961, while Neil Armstrong was the first to set foot on the moon on 21 July, 1969.

Since the 1960s, composite airframes and quieter, more efficient engines have become available, but the most important innovations have taken place in instrumentation and control. The arrival of solid-state electronics, the Global Positioning System, satellite communications, and increasingly small and powerful computers and LED displays, have dramatically changed the cockpits of airliners and, increasingly, of smaller aircraft as well. Pilots can navigate much more accurately and view terrain, obstructions, and other nearby aircraft on a map or through synthetic vision, even at night or in low visibility.

On June 21, 2004, Space Ship One became the first privately funded aircraft to make a spaceflight, opening the possibility of an aviation market outside the earth's atmosphere.

Civil Aviation

Civil aviation includes all non-military flying, both general aviation and scheduled air transport.

Lines Airbus A 330

While there were many more in the past, there are currently only five major manufacturers of civil transport aircraft:

- Airbus, based in Europe (primarily France).
- Boeing, based in the United States.
- Bombardier, based in Canada.
- Embraer, based in Brazil.
- Tupolev, based in Russia (scheduled to be merged into the United Aircraft Building Corporation).

Boeing, Airbus and Tupolev concentrate on larger airliners, while Bombardier and Embraer concentrate on commuter aircraft.

Until the 1970s, most major airlines were flag carriers, sponsored by their governments and heavily protected from competition. Since then, various open skies agreements have resulted in increased competition and choice for consumers, coupled with falling prices for airlines. The combination of high fuel prices, low fares, high salaries, and crises such as the September 11, 2001 attacks and the SARS epidemic have driven many older airlines to government-bailouts, bankruptcy or mergers. At the same time, low-cost carriers such as Ryanair and Southwest have flourished.

General Aviation

General aviation includes any flight that is not military and does not fly on a regular schedule, ranging from a recreational flight in a hang glider to a non-scheduled cargo flight in a Boeing 747. The majority of flights on any day will fall into this category.

Because of the huge range of activities, it is difficult to cover general aviation with a simple description—general aviation may include business flights, private aviation, flight training, ballooning, parachuting, gliding, hang gliding, aerial photography, foot-launched powered hang gliders, air ambulance, crop dusting, charter flights, traffic reporting, police air patrols, forest fire flighting, and many other types of flying.

Each country regulates aviation differently, but typically, general aviation falls under several different types of regulations depending on whether it is private or commercial and on the type of equipment involved.

Many small aircraft manufacturers, including Cessna, Piper, Diamond, Mooney, Cirrus Design, Raytheon, and others serve the general aviation market, with a focus on private aviation and flight training.

The most important recent developments for small aircraft have been the introduction of advanced avionics (including GPS) that were formerly found only in large airliners, and the introduction of composite materials to make small aircraft lighter and faster. Ultralight and homebuilt aircraft have also become increasingly popular for recreational use, since in most countries that allow private aviation, they are much less expensive and less heavily regulated than certified aircraft.

Simple airships were used as surveillance aircraft as early as the 18th century. Over the years, military aircraft have been built to meet ever increasing capability requirements. Manufacturers of military aircraft compete for contracts to supply their government's arsenal. Aircraft are selected based on factors like cost, performance, and the speed of production.

Military Aviation

Types of Military Aircraft

- Fighter aircraft's primary function is to destroy other aircraft (e.g. Sopwith Camel, A6M Zero, MiG-29).
- Ground attack aircraft are used against tactical earth-

bound targets (e.g. Junkers Stuka diver bomber, Ilyushin Il-2, and the A-10).

- Bombers are generally used against more strategic targets (e.g. Zeppelin, B-29 Superfortress, Tu-22, and the B-52).
- Surveillance aircraft have special capabilities used for reconnaissance (e.g. Rumpler Taube, de Havilland Mosquito, U-2 and MiG-25R).

Air Traffic Control (ATC)

Air Traffic Control (ATC) involves humans (typically on the ground) who communicate with aircraft to help maintain *separation*—that is, they ensure that aircraft are far enough apart horizontally or vertically that there is no risk of collision. Controllers may co-ordinate position reports provided by pilots, or in high traffic areas (such as the United States) they may use RADAR to see aircraft positions.

While the exact terminology varies from country to country, there are generally three different types of ATC:

- Control towers (including tower, ground control, clearance delivery and other services), which control aircraft within a small distance (typically 10-15 kms horizontal and 1,000 m vertical) of an airport.
- Terminal controllers, who control aircraft in a wider area (typically 50-80 kms) around busy airports.
- Centre controllers, who control aircraft enroute between airports.

ATC is especially important for aircraft flying under Instrument Flight Rules (IFR), where they may be in weather conditions that do not allow the pilots to see other aircraft. However, in very high-traffic areas, especially near major airports, aircraft flying under Visual Flight Rules (VFR) are also required to follow instructions from ATC.

In addition to separation from other aircraft, ATC may provide weather advisories, terrain separation, navigation assistance, and other services to pilots, depending on their workload.

It is important to note that ATC does not control all flights. The majority of VFR flights in North America are not required to talk to ATC at all (unless they are passing through a busy terminal area or using a major airport), and in many areas, such as northern Canada, ATC services are not available even for IFR flights at lower altitudes.

AVIATION SECTOR IN INDIA: AN INTRODUCTION

Revolutionised by liberalisation, the aviation sector in India has been marked by fast-paced change in the past few years. From being a service that few could afford, the sector has now graduated to being a fiercely competitive industry with the presence of a number of private and public airlines and several consumer-oriented offerings.

The market was galvanised a couple of years ago by the introduction of lower price tags which ensured that people could travel at the fraction of the original price of air travel. It was spurred further by the entry of Air Deccan, India's first budget airline, which offered hard-to-believe tariffs.

This was the trigger point for the sector to move from having simple economy, business and first class fares, to multiple slab tariffs such as apex fares, internet auctions, special discounts, bulk purchases and last day fares. Some of the tariffs offered are so low that they have brought airline fares neck-to-neck with upper-class railway fares. Little wonder then that the consumer prefers air travel to the railways.

The fare reduction has given the sector a huge boost with domestic and international traffic growing by 24.2 per cent and 18 per cent, respectively in April-December, 2005. Private airlines now account for 68.9 per cent of domestic traffic.

The boom in the aviation sector in India can be gauged by the fact that in one year, the number of people seeking pilot licences has multiplied three times. In April, 2005, it was 300. In April, 2006, the number rose to 1045.

In the last financial year, the civil aviation industry has achieved the biggest ever growth in aircraft movement and passenger traffic. Passenger traffic in the domestic airports increased by 22.3 per cent to 59.54 million, while aircraft movement increased by 14.2 per cent to 730,000.

Jet Airways tops the list of domestic and national carrier operators with 8,168 flights operating till June, 2005. Indian Airlines ranks second with 7,562 flights, followed by Sahara (3,225 flights), Air Deccan (2,889 flights), Spice Jet (483 flights) and Kingfisher Airlines (267 flights).

Aircraft manufacturer Airbus said, "India is one of the world's most promising markets and it is predicted that 100 million new urban middle-class consumers will become potential air travellers by 2010."

The increase in passenger traffic calls for upgraded infrastructure facilities. The international airport in Delhi and Mumbai are being modernised and upgraded through private sector participation. In the Joint Venture (JV), the Airports Authority of India (AAI) and other Government PSUs will hold 26 per cent equity. The balance 74 per cent will be held by the strategic partner. Foreign Direct Investment (FDI) in this transaction has been capped at 49 per cent.

In addition to these 10 non-metro airports, AAI has identified 15 more non-metro airports, namely, Agatti, Aurangabad, Bhopal, Bhubaneswar, Coimbatore, Indore, Khajuraho, Nagpur, Patna, Port Blair, Rajkot, Trichy, Vadodra, Varanasi and Vizag, for development.

Acquisitions

The India story dominated the Paris Air Show in mid-2005 with several existing and yet-to-be-launched airlines surprising the world with hefty order announcements. The virtually unknown low-cost start-up, IndiGo, lifted the tally of Indian deals to US$ 13 billion. The Delhi-based airline ordered a full fleet of 100 jets in the single-aisle A320 family.

- Low-cost carrier Spice Jet signed a US$ 700-million deal for ten aircraft with Boeing, even as Jet Airways and state-owned Air-India and Indian Airlines placed orders worth US$ 3 billion for aircraft engines.
- General Electric said it had won an order worth more than US$ 2.2 billion from Air-India for engines for the airline's new Boeing 777 and 787 fleets. The company said in a statement that Air-India had placed an order for the GE 90-115B engine for eight 777-200LRs and fifteen 777-300ERs. In addition, the airline has ordered for the GEnx engine to power twenty seven 787-8 aircraft.
- Indian, as Indian Airlines is called now, signed a deal worth US$ 500 million with CFM International to purchase engines for its newly acquired fleet of Airbus aircraft. The CFM 56-5b engine will power Indian's new fleet of 43 Airbus A 320s scheduled for delivery between late 2006 and 2010.

- The other deal at the world's largest air-show was an order for A 380 superjumbos and A 320s by Kingfisher Airlines.

Domestic air travel in India is predicted to grow 20 per cent over the next five years. Boeing has raised its 20-year market forecast for India for aircraft purchases from US$ 25 billion to US$ 35 billion. Both Airbus and Boeing are waiting for the next big order, expected from Air India. The airline is evaluating medium and large capacity aircraft and is expected to order 50 wide-body jets, worth almost US$ 5 billion at list prices.

The aviation sector is likely to see the launch of many new airlines, including Premier Airways, Star Air and East West Airlines this year. The first of these, Premier Airways, is formed by a group of ten NRIs in the US. The airline will be based in Chennai.

CIVIL AVIATION

Civil Aviation is one of two major categories of flying, representing all non-Military aviation, both private and commercial. Most of the countries in the world are members of the International Civil Aviation Organization (ICAO) and work together to establish common standards and recommended practices for civil aviation through that agency.

Civil aviation includes two major categories:

- *Scheduled air transport,* including all passenger and cargo flights operating on regularly-scheduled routes; and
- *General Aviation (GA),* including all other civil flights, private or commercial.

Although scheduled air transport is the larger operation in terms of passenger numbers, GA is larger in the number of flights. In the U.S., GA carries 166 million passengers each year, more than any individual airline, though less than all the airlines combined.

Civil Aviation Authorities

The Convention on International Civil Aviation (the *Chicago Convention*) was originally established in 1944: it states that signatories should collectively work to harmonize and standardize the use of airspace for safety, efficiency and regularity of air transport. All the States signatory to the Chicago Convention, now 188, are obliged to implement the Standards and Recommended Practices (SARPs) of the Convention.

Each signatory country has a Civil Aviation Authority (CAA) (such as the FAA in the United States) to oversee the following areas of civil aviation:

- *Personnel Licensing*: Regulating the basic training and issuance of licenses and certificates.
- *Flight Operations*: Carrying out safety oversight of commercial operators.
- *Airworthiness*: Issuing certificates of registration and certificates of airworthiness to civil aircraft, and overseeing the safety of maintenance organizations.
- *Aerodromes*: Designing and constructing aerodrome facilities.
- *Air Traffic Services*: Managing the traffic in a country's airspace.

AIRPORT

An airport is a facility where aircraft such as airplanes and helicopters can take off and land. An airport minimally consists of one runway or helipad (for helicopters), but other common components are hangars and terminal buildings.

Additionally, an airport may have a variety of facilities and infrastructure, including fixed base operator services, air traffic control, passenger facilities such as restaurants and lounges, and emergency services. A military airport is known as an airbase or air station. The terms **airfield, airstrip** and **aerodrome** may also be used to refer to airports of varying sizes. In some jurisdictions, the term airport is used where the facility is licensed as such by the relevant government organization (e.g. FAA, Transport Canada). Elsewhere the distinction is merely one of general appearance.

Attributes

Airports vary in size, with smaller or less-developed airports often having only a single runway shorter than 1,000 m (3,300 ft.). Larger airports for international flights generally have paved runways 2,000 m (6,600 ft.) or longer. Many small airports have dirt, grass, or gravel runways, rather than asphalt or concrete.

In the United States, the minimum dimensions for dry, hard landing fields are defined by the FAR Landing and Takeoff Field Lengths. These include considerations for safety margins during landing and takeoff. Typically, heavier aircraft require longer runways.

The longest public-use runway in the world is at Ulyanovsk Vostochny Airport, in Ulyanovsk, Russia. It has a length of 5,000 m (16,404 ft.). At 105 m (344 ft.), it should also be the world's widest paved runway.

As of 2006, there were approximately 49,000 airports around the world, including 14,858 in the United States alone.

Airport Ownership and Operation

Most of the world's airports are owned by local, regional, or national government bodies who then lease the airport to private corporations who oversee the airport's entire operation. For example, the British firm BAA plc operates seven of the commercial airports in the United Kingdom, as well as several other airports outside of the UK. Germany's Frankfurt International Airport is managed by the quasi-private firm Fraport AG.

In the United States, most airports are operated directly by government entities or government-created airport authorities (also known as port authorities). Only Indianapolis International Airport, which is owned by the City of Indianapolis and leased to BAA Indianapolis, Inc., a subsidiary of the British airport firm BAA plc, and Stewart International Airport in New York's Hudson Valley, are entirely operated by a private entity. Stewart International Airport was the nation's first privatized commercial airport and operates under a 99-year lease agreement with the New York State Department of Transportation. National Express Group, Inc., operates Stewart International Airport and is the United States subsidiary of the National Express Group, plc, in the United Kingdom.

Many U.S. airports still lease part or all of their facilities to outside firms, who operate specific functions such as retail management and parking. In the United States, all commercial airport runways are certified by the Federal Aviation Administration, but maintained by the local airport under the regulatory authority of the FAA.

Despite the reluctance to privatize airports in the United States (despite the FAA sponsoring a privatization programme since 1996), the Government-Owned, Contractor-Operated (GOCO) arrangement is the standard for the operation of commercial airports in the rest of the world.

In New Zealand, Auckland International Airport, the nation's main international airport, is fully privatised. Ownership and operation of the 1,497 hectare complex is vested entirely with Auckland International Airport Limited, a public company, with the only governmental involvement being Airways Corporation of New Zealand's operation of air traffic control systems. Similar arrangements pertain to Wellington and Christchurch airports, and most other main airports are operated by private companies.

Airport Structures

Airports are divided into *landside* and *airside* areas. Landside areas include parking lots, public transportation stations, tank farms and access roads. Airside areas include all areas accessible to aircraft, including runways, taxiways and ramps. Access from landside areas to airside areas is tightly controlled at most airports. Passengers on commercial flights access airside areas through terminals, where they can purchase tickets, clear security, check or claim luggage and board aircraft through gates. The waiting areas which provide passenger access to aircraft are typically called concourses, although this term is often used interchangeably with terminal.

The area where aircraft park next to a terminal to load passengers and baggage is known as a *ramp* (or, to the media and uninitiated, "the tarmac"). Parking areas for aircraft away from terminals are generally called aprons.

Both large and small airports can be towered or non-towered, depending on air traffic density and available funds. Due to their high capacity and busy airspace, most international airports have air traffic control located on site.

Airports with international flights have customs and immigration facilities. However, as some countries have agreements that allow travel between them without customs and immigrations, such facilities are not a definitive need for an international airport. International flights often require a more conspicuous level of physical security, although in recent years, many countries have adopted the same level of security for international and domestic travel.

Modern engineers and architects are developing "floating airports" which could be located several miles at sea and which would utilize designs such as Float Incorporated's Pneumatic Stabilized Platform (PSP) technology.

Shops and Food Services

Most international airports have shops and food courts. These services usually provide the passengers food and drinks before they board their flight. Many recognizable chain food restaurants have opened branches in large airports to serve often hungry passengers. International areas usually have a *duty free shop* ("Duty Free") where travellers are not required to pay the usual duty fees on items. Larger airlines often operate member-only lounges for premium passengers. London's Heathrow Airport, for example, is home to both a Harrods and a Hamleys Toy Shop, providing Duty Free for international passengers.

Airports have a captive audience, and consequently the prices charged for food is generally higher than are available elsewhere in the region. However, some airports, such as JFK's Terminal 8, guarantee prices comparable to those outside of the airport. Airport fees are fees commonly paid for use of services of airports, such as in the Subic Bay International Airport, known for charging airport fees. However, some airports now regulate food costs to keep them comparable to so-called "street prices". This term is a bit misleading as prices often match the MSRP but are almost never discounted.

Premium and VIP Services

Several mid-large size airports also have facilities for premium passengers. In the United States, these lounges are typically for international or long-haul first and business class passengers, paid members, and elite-level frequent fliers on long haul flights (regardless of what class they are in). In the rest of the world, the lounges are not open to purchase, but can be used by all premium passengers and most elite frequent fliers. Some lounges are comparatively spartan and only offer a quiet place to sit and work; other lounges include meals and massage services.

Cargo and Freight Services

In addition to people, airports are responsible for moving large volumes of cargo around the clock. Cargo airlines often have their own on-site and adjacent infrastructure to rapidly transfer parcels between ground and air modes of transportation.

Support Services

Aircraft maintenance, pilot services, aircraft rental, and hangar rental are most often performed by a Fixed Base Operator (FBO).

At major airports, particularly those used as hubs, airlines may operate their own support facilities.

Some airports, typically military airbases, have long runways used as emergency landing sites. Many airbases have arresting equipment for fast aircraft, known as Rotary Hydraulic Arrester Gear—a strong cable suspended just above the runway and attached to a hydraulic reduction gear mechanism. Together with the landing aircraft's arresting hook, it is used in situations where the brakes would have little or no effect.

Airport Access

Many large airports in the world are located next to or even above railway trunk routes, for instance Frankfurt International Airport, Amsterdam-Schiphol Airport, London Heathrow Airport, London Gatwick Airport and London Stansted. For local access, many airports have local train lines, rapid transit, light rail lines or other mass transit systems, for instance the AirTrain JFK at John F. Kennedy International Airport in New York and the Silver Line T at Boston's Logan International Airport by the MBTA (Massachusetts Bay Transportation Authority). These systems are usually directly connected to the main terminals. Large airports usually have access also through freeways from which cars fed into two access roads, designed as loops, one sitting on top of the other. One level is for departing passengers and the other is for arrivals. This road concept was pioneered at Los Angeles International Airport.

History and Development

The earliest airplane takeoff and landing sites were simply open, grassy fields. The plane could approach at any angle that provided a favourable wind direction. A slight improvement was the dirt-only field, which eliminated the drag from grass. However, these only functioned well in dry conditions. They would eventually be replaced by concrete surfaces that allowed all-weather landings in both daylight and at night.

Early airfields were often built for the purpose of entertainment. These aerodromes consisted of a grassy field, with a hangar for storage and servicing of airplanes, and observation stands for visitors. One of the world's first aerodromes was Taliedo Airport in Milan, which was opened in 1910 and in the mid-1920s was transformed into an airport and opened to commercial traffic.

It was closed in the 1930s because it was too small for the commercial transportation boom of those years. Linate Airport, which still serves as one of Milan's airports, was then built a few miles north. One of Europe's first airports was S. Darius and S. Girėnas airport (in Kaunas, Lithuania), which was opened in 1915.

The first known usage of the term "airport" appeared in a newspaper article in 1919, in reference to Bader Field in Atlantic City, New Jersey.

Increased aircraft traffic during World War I led to the construction of regular landing fields. Airplanes had to approach these from certain directions. This led to the development of aids for directing the approach and landing slope.

Following the war, some of these military airfields added commercial facilities for handling passenger traffic. One of the earliest such fields was Le Bourget, near Paris, The first international airport to open was the Croydon Airport, in South London. In 1922, the first permanent airport and commercial terminal solely for commercial aviation was built at Königsberg, Germany. The airports of this era used a paved "apron", which permitted night flying as well as landing heavier airplanes.

The first lighting used on an airport was during the later part of the 1920s; in the 1930s approach lighting came into use. These indicated the proper direction and angle of descent. The colours and flash intervals of these lights became standardized under the ICAO. In the 1940s, the slope-line approach system was introduced. This consisted of two rows of lights that formed a funnel indicating an aircraft's position on the glideslope. Additional lights indicated incorrect altitude and direction.

Following World War II, airport design began to become more sophisticated. Passenger buildings were being grouped together in an island, with runways arranged in groups about the terminal. This arrangement permitted expansion of the facilities. But it also meant that passengers had to travel further to reach their plane.

An improvement in the landing field was the introduction of grooves in the concrete surface. These ran perpendicular to the direction of the landing aircraft and served to draw off excess water in rainy conditions that could build up in front of the plane's wheels.

Airport construction boomed during the 1960s with the increase in jet aircraft traffic. Runways were extended out to 3 km (9800 feet). The fields were constructed out of reinforced concrete using a slip-form machine that produces a continual slab with no disruptions along the length.

Modern landing fields are thickest in the area where the plane touches down in order to support the force of the landing gear touching down. Runways are constructed as smooth and level as possible using measuring devices to correct for variations in height.

Airport Designation and Naming

Airports are uniquely represented by their IATA airport code and ICAO airport code. IATA airport codes are often abbreviated forms of the common name of the airport, such as PHL for Philadelphia International Airport. Airports sometimes retain their previous IATA code when an airport's name is changed. O'Hare International Airport in Chicago retains the IATA code ORD, from its former name of Orchard Field.

The name of the airport itself can be its location, such as Los Angeles International Airport. It can be the name of a prominent national celebrity, commonly a politician, e.g. John F. Kennedy International Airport, Ninoy Aquino International Airport, Indira Gandhi International Airport, Ataturk Airport or Charles de Gaulle International Airport. Airports may also be named after a person associated with the region it serves or prominent figures in aviation history, such as Will Rogers World Airport, Liverpool John Lennon Airport or Kingsford Smith International Airport (named for the Australian aviation pioneer), Sabiha Gökçen Airport (named for the first female combat pilot of the world, adoptive daughter of Mustafa Kemal Atatürk).

Airport Security

Airports are required to have safety precautions in most countries. Rules vary in different countries, but there are common elements worldwide. Airport security normally requires baggage checks, metal screenings of individual persons, and rules against any object that could be used as a weapon. Since the September 11, 2001 attacks, airport security has been dramatically increased worldwide.

Airport Operations

Outside the terminal, there is a large team of people who work in concert to ensure aircraft can land, take off, and move around quickly and safely. These processes are largely invisible to passengers, but they can be extraordinarily complex at large airports.

Air Traffic Control

The vast majority of the world's airports are non-towered, with no air traffic control presence at all. However, at particularly busy airports, or airports with other special requirements, there is an Air Traffic Control (ATC) system whereby controllers (usually ground-based) direct aircraft movements via radio or other communications links. This coordinated oversight facilitates safety and speed in complex operations where traffic moves in all three dimensions. Air traffic control responsibilities at airports are usually divided into at least two main areas: *ground* and *tower*, though a single controller may work both stations. The busiest airports also have *clearance delivery*, *apron control*, and other specialized ATC stations.

Ground Control is responsible for directing all ground traffic in designated "movement areas," except the traffic on runways. This includes planes, baggage trains, snowplows, grass cutters, fuel trucks, and a wide array of other vehicles. Ground Control will instruct these vehicles on which taxiways to use, which runway they will use (in the case of planes), where they will park, and when it is safe to cross runways. When a plane is ready to take off it will stop short of the runway, at which point it will be turned over to Tower Control. After a plane has landed, it will depart the runway and be returned to Ground Control.

Tower Control aircraft on the runway and in the controlled airspace immediately surrounding the airport. Tower controllers may use radar to identify and accurately locate an aircraft's position in three-dimensional space, or they may rely on pilot position reports and visual observation. They coordinate the sequencing of aircraft in the traffic pattern and direct aircraft on how to safely join and leave the circuit. Aircraft which are only passing through the airspace must also contact Tower Control in order to be sure that they remain clear of other traffic and do not disrupt operations.

Traffic Pattern

All airports use a traffic pattern (often called a *traffic circuit* outside the U.S.) to assure smooth traffic flow between departing and arriving aircraft. Generally, this pattern is a circuit consisting of five "legs" that form a rectangle (two legs and the runway form one side, with the remaining legs forming three more sides). Each leg is named, and ATC directs pilots on how to join and leave the circuit. Traffic patterns are flown at one specific altitude, usually 800 or 1000 ft. Above Ground Level (AGL). Standard traffic patterns are *left-handed,* meaning all turns are made to the left. Right-handed patterns do exist, usually because of obstacles such as a mountain, or to reduce noise for local residents. The predetermined circuit helps traffic flow smoothly because all pilots know what to expect, and helps reduce the chance of a mid-air collision.

At extremely large airports, a circuit is in place but not usually used. Rather, aircraft (usually only commercial with long routes) request approach clearance while they are still hours away from the airport, often before they even take off from their departure point. Large airports have a frequency called *Clearance Delivery* which is used by departing aircraft specifically for this purpose. This then allows airplanes to take the most direct approach path to the runway and land without worrying about interference from other aircraft. While this system keeps the airspace free and is simpler for pilots, it requires detailed knowledge of how aircraft are planning to use the airport ahead of time and is therefore only possible with large commercial airliners on pre-scheduled flights. The system has recently become so advanced that controllers can predict whether an aircraft will be delayed on landing before it even takes off; that aircraft can then be delayed on the ground, rather than wasting expensive fuel waiting in the air.

Navigational Aids

Before takeoff, pilots usually check an Automatic Terminal Information Service (ATIS) for information about airport conditions where they exist. The ATIS contains information about weather, which runway and traffic patterns are in use, and other information that pilots should be aware of.

When flying, there are a number of aids available to pilots, though not all airports are equipped with them. A Visual Approach Slope Indicator (VASI) helps pilots fly a perfect approach for landing once they have found the airport. Some airports are equipped with a VHF omnidirectional range (VOR) to help pilots find the direction to the airport. VORs are often accompanied by a Distance Measuring Equipment (DME) to determine the distance to the VOR. VORs are also located off airports, where they serve to provide airways for aircraft to navigate upon. In poor weather, pilots will use an Instrument Landing System to find the runway and fly the correct approach, even if they cannot see the ground. The number of instrument approaches based on the use of the Global Positioning System (GPS) is rapidly increasing and may eventually be the primary means for instrument landings.

Larger airports sometimes offer Precision Approach Radar (PAR), but these systems are more common at military air bases than civilian airports. The aircraft's horizontal and vertical movement is tracked via radar, and the controller tells the pilot his position relative to the approach slope. Once the pilots can see the runway lights, they may continue with a visual landing.

Guidance Signs

Airport guidance signs provide direction and information to taxiing aircraft and airport vehicles and assist in safe and expedient movement of aircraft. Smaller airports may have few or no signs, relying instead on airport diagrams and charts.

There are two classes of signage at airports, with several types of each:

Operational Guidance Signs

- *Location Signs*: Yellow on black background. Identifies the runway or taxiway currently on or entering.
- *Direction/Runway Exit Signs*: Black on yellow. Identifies the intersecting taxiways the aircraft is approaching, with an arrow indicating the direction to turn.
- *Other*: Many airports use conventional traffic signs such as stop and yield signs throughout the airport.

Mandatory Instruction Signs

Mandatory instruction signs are white on red. They show entrances to runways or critical areas. Vehicles and aircraft are required to stop at these signs until the control tower gives clearance to proceed.

- *Runway Signs*: White on a red. These signs simply identify a runway intersection ahead.
- *Frequency Change Signs*: Usually a stop sign and an instruction to change to another frequency. These signs are used at airports with different areas of ground control.

- *Holding Position Signs*: A single solid yellow bar across a taxiway indicates a position where ground control may require a stop. If a two solid yellow bars and two dashed yellow bars are encountered, this indicates a holding position for a runway intersection ahead; runway holding lines must never be crossed without permission. At some airports, a line of red lights across a taxiway is used during low visibility operations to indicate holding positions.

Lighting

Many airports have lighting that help guide planes using the runways and taxiways at night or in rain or fog.

On runways, green lights indicate the beginning of the runway for landing, while red lights indicate the end of the runway. Runway edge lighting is white lights spaced out on both sides of the runway, indicating the edge. Some airports have more complicated lighting on the runways including lights that run down the centerline of the runway and lights that help indicate the approach (an Approach Lighting System, or ALS). Low-traffic airports may use Pilot Controlled Lighting to save electricity and staffing costs.

Along taxiways, blue lights indicate the taxiway's edge, and some airports have embedded green lights that indicate the centerline.

Wind Indicators

Planes take-off and land *into* the wind in order to achieve maximum performance. Wind speed and direction information is available through the ATIS or ATC, but pilots need instantaneous information during landing. For this purpose, a windsock is kept in view of the runway.

Safety Management

Air safety is an important concern in the operation of an airport, and almost every airfield includes equipment and procedures for handling emergency situations. Commercial airfields include one or more emergency vehicles and their crew that are specially equipped for dealing with airfield accidents, crew and passenger extractions, and the hazards of highly flammable airplane fuel. The crews are also trained to deal with situations such as bomb threats, hijacking and terrorist activities.

Potential airfield hazards to aircraft include debris, nesting birds, and environmental conditions such as ice or snow. The fields must be kept clear of debris using cleaning equipment so that loose material doesn't become a projectile and enter an engine duct. In adverse weather conditions, ice and snow clearing equipment can be used to improve traction on the landing strip. For waiting aircraft, equipment is used to spray special deciding fluids on the wings.

Many airports are built near open fields or wetlands. These tend to attract bird populations, which can pose a hazard to aircraft in the form of bird strikes. Airport crews often need to discourage birds from taking up residence.

Depending on their configuration, an airport can have sections of their pavement where collisions between airplanes on the ground can tend to occur. Records are kept of any incursions where airplanes or vehicles are in an inappropriate location, allowing these "hot spots" to be identified. These locations then undergo special attention by transportation authorities (such as the FAA in the US) and airport administrators in order to reduce accidents.

During the 1980s, a phenomenon known as microburst became a growing concern due to accidents caused by microburst wind shear. (For example, see Delta Air Lines Flight 191.) Microburst radar was developed as an aid to safety during landing, giving two to five minutes warning to aircraft in the vicinity of the field of a microburst event.

Some airfields now have a special surface material at the end of the runway that behaves somewhat like styrofoam, bringing the plane to a relatively rapid halt as the material disintegrates. These surfaces are useful when the runway is located next to a body of water or other hazard, and prevent the planes from overrunning the end of the field.

Environmental Concerns

The traffic generated by airports both in the air and on the surface can be a major source of aviation noise and air pollution which may interrupt nearby residents' sleep and produce other noise health effects. The construction of new airports, or addition of runways to existing airports, is often resisted by local residents because of the effect on the countryside, historical sites, local flora and fauna. As well, due to the risk of collision between birds and airplanes, large airports undertake population control programmes where they frighten or shoot birds to ensure the safety of air travellers.

The construction of airports has been known to change local weather patterns. For example, because they often flatten out large areas, they can be susceptible to fog in areas where fog rarely forms. In addition, because they generally replace trees and grass with pavement, they often change drainage patterns in agricultural areas, leading to more flooding, run-off and erosion in the surrounding land.

Military Airbase

An airbase, sometimes referred to as a *military airport* or *airfield,* provides basing and support of military aircraft. Some airbases provide facilities similar to their civilian counterparts. For example, RAF Brize Norton in Oxfordshire, England has a terminal which caters to passengers for the Royal Air Force's scheduled Tristar flights to the Falkland Islands. Military airbases may also be co-located with civilian airports, sharing the same tower/air traffic control facilities, runways, taxiways and emergency services, but with separate terminals, parking areas, hangars and shelter areas. Examples of this are Bardufoss Airport/Bardufoss Air Station and Gardermoen Airport/Gardermoen Air Station, both in Norway. A special variant of a military airfield is the Aircraft Carrier.

Aircraft Carriers

An aircraft carrier is a warship that functions as a floating airport for military aircraft. Aircraft carriers allow a naval force to project air power great distances without having to depend on local bases for land-based aircraft. After their development in First World War, aircraft carriers replaced the battleship as the centrepiece of a modern fleet during Second World War. Unescorted carriers are considered vulnerable to missile or submarine attacks and therefore travel as part of a *carrier battle group* that includes a wide array of other ships with specific functions.

Airports in Entertainment

Airports have occasionally played major roles in motion pictures and television shows due to being transportation hubs, but also because of their unique characteristics. One such example of this is the movie *The Terminal,* a film about a man who becomes permanently grounded in an airport terminal and must survive only on the food and shelter provided by the airport. Movies such as *Airplane!, Airport, Die Hard II, Jackie Brown, Get Shorty, Unaccompanied Minors* and *The Langoliers,* as well as television series like *Lost* also have significant parts of their story set within airports.

Also, several computer simulation games were created in which the player is in charge of an airport. Names include Airport Tycoon and the sequels; Airport Tycoon 2 and Airport Tycoon 3.

Airport Directories

Each national aviation authority has its own system for pilots to be able to keep track of information about airports in their country.

- The United States uses the **Airport/Facility Directory (A/FD)**, seven volumes that contain information such as elevation, airport lighting, runway information, communications, hours of operation, nearby NAVAIDs and much more.
- In Canada, a single publication, the **Canada Flight Supplement (CFS)** provides equivalent information.
- The most comprehensive, consumer/business directory of commercial airports in the world (primarily for airports as businesses, rather than for pilots) is organized by the trade group Airports Council International.

AIR TRAFFIC CONTROL

Air Traffic Control (ATC) is a service provided by ground-based controllers who direct aircraft on the ground and in the air. A controller's primary task is to *separate* certain aircraft—to prevent them from coming too close to each other by use of lateral, vertical and longitudinal separation. Secondary tasks include ensuring orderly and expeditious flow of traffic and providing information to pilots, such as weather, navigation information and NOTAMs (Notices to Airmen).

In many countries, ATC services are provided throughout the majority of airspace, and its services are available to all users (private, military and commercial). When controllers are responsible for separating some or all aircraft, such airspace is called "controlled airspace" in contrast to "uncontrolled airspace" where aircraft may fly without the use of the air traffic control system. Depending on the type of flight and the class of airspace,

ATC may issue *instructions* that pilots are required to follow, or merely *flight information* (in some countries known as *advisories*) to assist pilots operating in the airspace. In all cases, however, the pilot in command has final responsibility for the safety of the flight, and may deviate from ATC instructions in an emergency.

Air traffic control services can be divided into two major sub-specialties, **terminal** or **airport control** and **en-route** or **area control.**

Terminal or Airport control includes the control of traffic (aircraft and vehicles) on the airport surface and airborne aircraft within the immediate airport environment. Generally, this is approximately a 30 to 50 nautical mile (56 to 93 km) radius of the airport, from the surface to about 10,000 ft. (about 3,050 m). Terminal controllers work in facilities called control towers and terminal control centers (called *Terminal Radar Approach Control,* or TRACON, facilities in the U.S.). At some locations, controllers are shared between tower control and the terminal control center, while at others the tower and the terminal control center are completely separate entities. For example, Philadelphia International Airport is served by a combined ("up/down") facility, while Chicago's O'Hare Airport is served by a control tower at the airport, and a remote TRACON located at Elgin, Illinois.

En-route controllers, also called Center or Area controllers, control the traffic between the terminals. They can also control traffic in and out of airports where the traffic volume does not warrant the establishment of a terminal ATC operation or during periods when a terminal operation is closed (e.g., mid-night to 6:00 am). En-route controllers work at facilities called Area Control Centers or Air Route Traffic Control Centers (ARTCCs).

Terminal Control

The primary method of controlling the immediate airport environment is visual observation from the control tower. The tower is a tall, windowed structure located on the airport grounds. **Aerodrome** or **Tower** controllers are responsible for the separation and efficient movement of aircraft and vehicles operating on the taxiways and runways of the airport itself, and aircraft in the air near the airport, generally 2 to 5 nautical miles (4 to 9 km) depending on the airport procedures.

Radar displays are also available to controllers at some airports. Controllers may use a radar system called Secondary Surveillance Radar also known as Airport Surveillance Radar for airborne traffic approaching and departing. These displays include a map of the area, the position of various aircraft, and data tags that include aircraft identification, speed, heading, and other information described in local procedures.

The areas of responsibility for tower controllers fall into three general operational disciplines; **Ground Control**, **Local** or **Air Control** and **Clearance Delivery**—other categories, such as **Apron Control** or **Ground Movement Planner**, may exist at extremely busy airports. While each tower's procedures will vary and while there may be multiple teams in larger towers that control multiple runways, the following provides a general concept of the delegation of responsibilities within the tower environment.

Ground Control

Ground Control (sometimes known as *Ground Movement Control* abbreviated to *GMC* or *Surface Movement Control* abbreviated to *SMC*) is responsible for the airport "maneuvering" areas, or areas not released to the airlines or other users. This generally includes all taxiways, holding areas, and some transitional aprons or intersections where aircraft arrive having vacated the runway and departure gates. Exact areas and control responsibilities are clearly defined in local documents and agreements at each airport. Any aircraft, vehicle, or person walking or working in these areas is required to have clearance from the ground controller. This is normally done via VHF radio, but there may be special cases where other processes are used. Most aircraft and airside vehicles have radios. Aircraft or vehicles without radios will communicate with the tower via aviation light signals or will be led by vehicles with radios. People working on the airport surface normally have a communications link through which they can reach or be reached by ground control, commonly either by handheld radio or even cell phone. Ground control is vital to the smooth operation of the airport because this position might constrain the order in which the aircraft will be sequenced to depart, which can affect the safety and efficiency of the airport's operation.

Some busier airports have Surface Movement Radar (SMR), such as, ASDE-3, AMASS or ASDE-X, designed to display aircraft and vehicles on the ground. These are used by the ground controller as an additional tool to control ground traffic, particularly at night or in poor visibility. There are a wide range of capabilities on these systems as they are being modernized. Older systems will display a map of the airport and the target. Newer systems include the capability to display higher quality mapping, radar target, data blocks and safety alerts.

Local or Air Control

Local or Air Control (most often referred to as the generic "Tower" control, although Tower control can also refer to a combination of the local, ground and clearance delivery positions) is responsible for the active runway surfaces. The Air Traffic Control Tower clears aircraft for take off or landing and ensures the runway is clear for these aircraft. To accomplish this, Tower controllers are normally given 2 to 5 nautical miles (4 to 9 km) of airspace around the airport, allowing them to give the clearances necessary for airport safety. If the tower controller detects any unsafe condition, a landing aircraft will be told to "go-around" and will be re-sequenced into the landing pattern by the approach or terminal area controller.

Within the tower, a highly disciplined communications process between tower and ground control is an absolute necessity. Ground control must request and gain approval from tower control to cross any runway with any aircraft or vehicle. Likewise, tower control must ensure ground control is aware of any operations that impact the taxiways and must work with the approach radar controllers to ensure "holes" or "gaps" in the arrival traffic are created (where necessary) to allow taxiing traffic to cross runways and to allow departures aircraft to take off. Crew Resource Management procedures are often used to ensure this communication process is efficient and clear.

Clearance Delivery

Clearance delivery is the position that issues route clearances to aircraft before they commece taxiing. These contain details of the route that the aircraft is expected to fly after departure. This position will, if necessary, coordinate with the en-route center and national command center or flow control to obtain releases for aircraft. Often however such releases are given automatically or by controlled by local agreements allowing "free-flow" departures. When weather or extremely high demand for a certain airports or airspace become a factor, there may be ground "stops" (or "slot delays"), the airport may go "IFR", or re-routes may be necessary to ensure the system does not get overloaded. The primary responsibility of the clearance delivery position is to ensure that the aircraft have the proper route and release time ("slot time"). This information is also coordinated with the en-route center and the ground controller in order to ensure the aircraft reaches the runway in time to meet the release time provided by the command center. At some airports the clearance delivery controller also plans aircraft pushbacks and engine starts and is known as Ground Movement Planner (GMP): this position is particularly important at heavily congested airports to prevent taxiway and apron gridlock.

Approach and Terminal Control

Many airports have a radar control facility that is associated with the airport. In most countries, this is referred to as *Approach* or *Terminal Control;* in the U.S., it is often still referred to as a TRACON or Terminal Radar Approach Control facility. While every airport varies, terminal controllers usually handle traffic in a 30 to 50 nautical mile (56 to 187 kms) radius from the airport and from the surface up to 10,000 feet. Where there are many busy airports in close proximity, one single terminal control may service all the airports. The actual airspace boundaries and altitudes assigned to a terminal control are based on factors such as traffic flows, neighbouring airports and terrain, and vary widely from airport to airport: a large and complex example is the London Terminal Control Center which controls traffic for 5 main London airports up to 20,000 feet and out to 100+ nautical miles.

Terminal controllers are responsible for providing all ATC services within their airspace. Traffic flow is broadly divided into departures, arrivals, overflights and VFR aircraft. As aircraft move in and out of the terminal airspace, they are handed off to the next appropriate control facility (a control tower, an en-route control facility, or a bordering terminal or approach control). Terminal control is responsible for ensuring that aircraft are at an appropriate altitude when they are handed off, and that aircraft arrive at a suitable rate for landing.

Not all airports have a radar approach or terminal control available. In this case, the en-route center or a neighbouring terminal or approach control will co-ordinate directly with the tower on the airport and vector inbound aircraft to a position from where they can land visually. At some of these airports, the tower may provide a non-radar procedural approach service to arriving aircraft handed over from a radar unit before they are visual to land.

En-route, Center or Area Control

ATC provides services to aircraft in flight between airports as well. The level of service is dependent on the type of flight the aircraft falls under (IFR or VFR), the type of airspace the aircraft is in and the services requested by the pilots.

En-route air traffic controllers issue clearances and instructions for airborne aircraft, and pilots are required to comply with these instructions. En-route controllers also provide air traffic control services to many smaller airports around the country, including clearance off of the ground and clearance for approach to an airport. Controllers adhere to a set of separation standards that define the minimum distance allowed between aircraft. These distances vary depending on the equipment and procedures used in providing ATC services.

Pilots fly under one of two sets of rules for separation: Visual Flight Rules (VFR) or Instrument Flight Rules (IFR). Air traffic controllers have different responsibilities to aircraft operating under the different sets of rules.

General Characteristics

En-route air traffic controllers work in facilities called Area Control Centers, each of which is commonly referred to as a "Center". The United States uses the equivalent term Air Route Traffic Control Center (ARTCC). Each center is responsible for many thousands of square miles of airspace (known as a Flight Information Region) and for the airports within that airspace. Centers control IFR aircraft from the time they depart an airport or terminal area's airspace to the time they arrive at another airport or terminal area's airspace. Centers may also "pick up" VFR aircraft that are already airborne and integrate them into the IFR system. These aircraft must, however, remain VFR until the Center provides a clearance.

Center controllers are responsible for climbing the aircraft to their requested altitude while, at the same time, ensuring that the aircraft is properly separated from all other aircraft in the immediate area. Additionally, the aircraft must be placed in a flow consistent with the aircraft's route of flight. This effort is complicated by crossing traffic, severe weather, special missions that require large airspace allocations, and traffic density. When the aircraft approaches its destination, the center is responsible for meeting altitude restrictions by specific points, as well as providing many destination airports with a traffic flow, which prohibits all of the arrivals being "bunched together". These "flow restrictions" often begin in the middle of the route, as controllers will position aircraft landing in the same destination so that when the aircraft are close to their destination they are sequenced.

As an aircraft reaches the boundary of a Center's control area it is "handed off" or "handed over" to the next Area Control Center. In some cases this "hand-off" process involves a transfer of identification and details between controllers so that air traffic control services can be provided in a seamless manner; in other cases local agreements may allow "silent handovers" such that the receiving center does not require any co-ordination if traffic is presented in an agreed manner. After the hand-off, the aircraft is given a frequency change and begins talking to the next controller. This process continues until the aircraft is handed off to a terminal controller ("approach").

Radar Coverage

Since centers control a large airspace area, they will typically use long range radar that has the capability to see aircraft within 200 nautical miles (370 km) of the radar antenna. They may also use TRACON radar data to control when it provides a better "picture" of the traffic or when it can fill in a portion of the area not covered by the long range radar.

In the U.S. system, at higher altitudes, over 90 per cent of the U.S. airspace is covered by radar and often by multiple radar systems; however, coverage may be inconsistent at lower altitudes used by unpressurized aircraft due to high terrain or distance from radar facilities. A center may require numerous radar systems to cover the airspace assigned to them, and may also rely on pilot position reports from aircraft flying below the floor of radar coverage. This results in a large amount of data being available to the controller. To address this, automation systems have been designed that consolidate the radar data for the controller. This consolidation includes eliminating duplicate radar returns, ensuring the best radar for each geographical area is providing the data, and displaying the data in an effective format.

Centers also exercise control over traffic travelling over the world's ocean areas. These areas are also FIRs. Due to the fact that there are no radar systems available for oceanic control, oceanic controllers provide ATC services using procedural control. These procedures use aircraft position reports, time, altitude, distance, and speed to ensure separation. Controllers record information on flight progress strips and in specially developed oceanic computer systems as aircraft report positions. This process requires that aircraft be separated by greater distances, which reduces the overall capacity for any given route.

Some Air Navigation Service Providers (e.g Airservices Australia, Alaska Center, etc.) are implementing Automatic Dependent Surveillance—Broadcast (ADS-B) as part of their surveillance capability. This new technology reverses the radar concept. Instead of radar "finding" a target by interrogating the transponder, ADS transmits the aircraft's position several times a second. ADS also has other modes such as the "contract" mode where the aircraft reports a position based on a predetermined time interval. This is significant because it can be used where it is not possible to locate the infrastructure for a radar system (e.g. over water). Computerized radar displays are now being designed to accept ADS inputs as part of the display. As this technology develops, oceanic ATC procedures will be modernized to take advantage of the benefits this technology provides.

Problems

Traffic

The day-to-day problems faced by the air traffic control system are primarily related to the volume of air traffic demand placed on the system, and weather. Several factors dictate the amount of traffic that can land at an airport in a given amount of time. Each landing aircraft must touch down, slow, and exit the runway before the next crosses the end of the runway. This process requires at least one and up to four minutes for each aircraft. Allowing for departures between arrivals, each runway can thus handle about 30 arrivals per hour. A typical large airport with two arrival runways can handle about 60 arrivals per hour in good weather. Problems begin when airlines schedule more arrivals into an airport than can be physically handled, or when delays elsewhere cause groups of aircraft that would otherwise be separated in time to arrive simultaneously. Aircraft must then be delayed in the air by holding over specified locations until they may be safely sequenced to the runway. Up until the 1990s, holding, which has significant environmental and cost implications, was a routine occurrence at many airports. Advances in computers now allow the sequencing of planes hours in advance. Thus, planes may be delayed before they even take off (by being given a "slot"), or may reduce power in flight and proceed more slowly thus significantly reducing the amount of holding.

Weather

Beyond runway capacity issues, weather is a major factor in traffic capacity. Rain or ice and snow on the runway cause landing aircraft to take longer to slow and exit, thus reducing the safe arrival rate and requiring more space between landing aircraft. This, in turn, increases airborne delay for holding aircraft. If more aircraft are scheduled than can be safely and efficiently held in the air, a ground delay programme may be established, delaying aircraft on the ground before departure due to conditions at the arrival airport.

In ACCs, a major weather problem is thunderstorms, which present a variety of hazards to aircraft. Aircraft will deviate around storms, reducing the capacity of the en-route system by requiring more space per aircraft, or causing congestion as many aircraft try to move through a single hole in a line of thunderstorms. Occasionally weather considerations cause delays to aircraft prior to their departure as routes are closed by thunderstorms.

Much money has been spent on creating software to streamline this process. However, at some Area Control Centers, air traffic controllers still record data for each flight on strips of paper and personally coordinate their paths. In newer sites, these flight progress strips have been replaced by electronic data presented on computer screens. As new equipment is brought in, more and more sites are upgrading away from paper flight strips.

Call Signs

A prerequisite to safe air traffic separation is the assignment and use of distinctive call signs. These are permanently allocated by ICAO on request usually to scheduled flights and some air forces for military flights. They are written call signs with 3-letter combination like KLM, AAL, SWA, BAW, DLH followed by the flight number, like AAL872, BAW018. As such they appear on flight plans and ATC radar labels. There are also the *audio* or *Radio-telephony* call signs used on the radio contact between pilots and Air Traffic Control not always identical with the written ones. For example, BAW stands for British Airways but on the radio you will only hear the word *Speedbird* instead. By default, the call sign for any other flight is the registration number (tail number) of the aircraft, such as "N12345" or "CG-ABC". The term *tail number* is due to the fact that a registration number is usually painted somewhere on the tail of a plane, yet this is not a rule. Registration numbers may appear on the engines, anywhere on the fuselage, and often on the wings. The short *Radio-telephony* call signs for these tail numbers is the first letter followed by the last two, like C-BC spoken as Charlie-Bravo-Charlie for CG-ABC or the last 3 letters only like ABC spoken Alpha-Bravo-Charlie for CG-ABC or the last 3 numbers like 345 spoken as tree-fower-fife for N12345.

The flight number part is decided by the aircraft operator. In this arrangement, an identical call sign might well be used for the same scheduled journey each day it is operated, even if the departure time varies a little across different days of the week. The call sign of the return flight often differs only by the final digit from the outbound flight. Generally, airline flight numbers are even if eastbound, and odd if westbound. In order to reduce the possibility of two call signs on one frequency at any time sounding too similar, a number of airlines, particularly in Europe, have started using alphanumeric call signs that are not based on flight numbers. For example DLH23LG, spoken as lufthansa-too-tree-lima-golf. Additionally it is the right of the air traffic controller to change the 'audio' call sign for the period the flight is in his sector if there is a risk of confusion, usually choosing the tail number instead.

Before around 1980 IATA and ICAO were using the same 2-letter call signs. Due to the larger number of new airlines after deregulation ICAO established the 3-letter call signs as mentioned above. The IATA call signs are currently used in aerodromes on the announcement tables but never used any longer in Air Traffic Control. For example, AA is the IATA call sign for the ICAO—ATC equivalent AAL. Other examples include LY/ELY for El Al, DL/DAL for Delta Air Lines, LH/DLH for Lufthansa etc.

Technology

Many technologies are used in air traffic control systems. Primary and secondary radar are used to enhance a controller's "situational awareness" within his assigned airspace—all types of aircraft send back primary echoes of varying sizes to controllers' screens as radar energy is bounced off their skins, and transponder-equipped aircraft reply to secondary radar interrogations by giving an ID (Mode A), an altitude (Mode C) and/or a unique call sign (Mode S). Certain types of weather may also register on the radar screen.

These inputs, added to data from other radars, are correlated to build the air situation. Some basic processing occurs on the radar tracks, such as calculating ground speed and magnetic headings.

Other correlations with electronic flight plans are also available to controllers on modern operational display systems.

Some tools are available in different domains to help the controller further:

- *Conflict Alert* (*CA*): A tool that checks possible conflicting trajectories and alerts the controller. The most common used is the STCA (Short Term CA) that is activated about 2 minutes prior the loss of separation. The algorithms used may also provide in some systems a possible vectoring solution, that is, the way to turn or descend/climb the aircraft in order to avoid infringing the minimum safety distance or altitude clearance.
- *Minimum Safe Altitude Warning* (*MSAW*): A tool that alerts the controller if an aircraft appears to be flying too low to the ground or will impact terrain based on its current altitude and heading.
- *System Coordination* (*SYSCO*) to enable controller to negotiate the release of flights from one sector to another.
- *Area Penetration Warning* (*APW*) to inform a controller that a flight will penetrate a restricted area.
- *Arrival and Departure* manager to help sequence the takeoff and landing of aircraft.
- *Converging Runway Display Aid* (*CRDA*) enables Approach Controllers to run two final approaches that intersect and make sure that go arounds are minimized.
- *Final Approach Spacing Tool* (*FAST*) gives aircraft a runway assignment that the Approach Controller will give to the aircraft. FAST can also suggest vectors for downwind and

base with the correct timing. In Europe the equivalent system is known under the term *metering system* and predicts the future spacing of approaching aircraft on the runway.

- *User Request Evaluation Tool* (*URET*) takes paper strips out of the equation for En-Route controllers at ARTCCs. By providing a display that shows all aircraft in or entering the sector. URET provides conflict resolution up to 30 minutes in advance.
- *Mode S*: Provides a data downlink of flight parameters via Secondary Surveillance Radars allowing radar processing systems and therefore controllers to see various data on a flight, including airframe unique id, indicated airspeed and flight director selected level, amongst others.
- *Controller Pilot Datalink Communications (CPDLC):* Allows digital messages to be sent between controllers and pilots, avoiding the need to use radiotelephony. It is especially useful in areas where difficult-to-use HF radiotelephony was previously used for communication with aircraft, e.g oceans. This is currently in use in various parts of the world including the Atlantic and Pacific oceans.
- *Automatic Dependent Surveillance (ADS-B)*: Provides a data downlink of various flight parameters to air traffic control systems via the Transponder (1090 MHz) and reception of those data by other aircraft in the vicinity. The most important is the aircraft's latitude, longitude and level: such data can be utilized to create a radar-like display of aircraft for controllers and thus allows a form of pseudo-radar control to be done in areas where the installation of radar is either prohibitive on the grounds of low traffic levels, or technically not feasible (e.g. oceans). This is currently in use in Australia and parts of the Pacific Ocean and Alaska.
- *The Extended Computer Display System (EXCDS)*: A system of electronic flight strips replacing the old paper strips. EXCDS allows controllers to manage electronic flight data online using touch-sensitive display screens resulting in fewer manual functions and a greater focus on safety. The system has also been sold to the Air Navigation Services Providers in the United Kingdom and Denmark.

Major Accidents

Failures in the system have caused delays or even, in rare cases, crashes. On 1 July, 2002 a Tupolev Tu-154 and Boeing 757 collided above Überlingen near the boundary between German and Swiss-controlled airspace when a Skyguide-employed controller apparently gave instructions to the southbound Tupolev to descend despite an instruction from the on-board automatic Traffic Collision Avoidance System software to climb. The northbound Boeing, equipped with similar avionics, was already descending due to a software prompt. All passengers and crew died in the resultant collision. Skyguide company publicity had previously acknowledged that the relatively small size of Swiss airspace makes real-time cross-boundary liaison with adjoining authorities particularly important. See *Bashkirian Airlines Flight 2937* for more on this accident. It is worth noting that currently air traffic controllers have no way of knowing if or when the TCAS system is issuing resolution advisories to pilots. They also do not know what the advisory is telling the pilots. Therefore, pilots are supposed to immediately report TCAS resolution advisories and follow them as soon as possible. Consequently, they should ignore ATC instructions until they have reported to the ground that they are clear of the conflict.

Other fatal collisions between airliners have occurred over India and Yugoslavia. When a risk of collision is identified by aircrew or ground controllers an "air miss" or "air prox" (air proximity) report can be filed with the air traffic control authority concerned. The worst fatal collision between airliners actually took place on the ground, on 27 March, 1977, in what is known as the Tenerife disaster.

The FAA has spent over USD$3 billion on software, but a fully-automated system is still over the horizon. In 2002 the UK brought a new area control centre into service at Swanwick, in Hampshire, relieving a busy suburban centre at West Drayton in Middlesex, north of London Heathrow Airport. Software from Lockheed-Martin predominates at Swanwick. The Swanwick facility, however, has been troubled by software and communications problems causing delays and occasional shutdowns, paralyzing air traffic in the area.

Air Navigation Service Providers (*ANSPs*)

In the United States, the Federal Aviation Administration (FAA) provides this service to all aircraft in the National Airspace System

(NAS). With the exception of facilities operated by the Department of Defence (DoD), the FAA is responsible for all aspects of U.S. Air Traffic Control including hiring and training controllers, although there are contract towers located in many parts of the country. DoD facilities are generally staffed by military personnel and operate separately but concurrently with FAA facilities, under similar rules and procedures. A contract tower is an Airport Traffic Control Tower (ATCT) that performs the same function as an FAA-run ATCT but is staffed by employees of a private company (Martin State Airport in Maryland is an example). In Canada, Air Traffic Control is provided by NAV CANADA, a private, non-share capital corporation that operates Canada's civil air navigation service.

- Austria—Austro Control.
- Australia—Airservices Australia (State Owned Corporation) and the Royal Australian Air Force.
- Belgium—Belgocontrol.
- Bulgaria—Air Traffic Services Authority.
- Canada—NAV CANADA—formerly provided by Transport Canada.
- Denmark—Naviair (Danish ATC).
- Europe—Eurocontrol (European Organisation for the Safety of Air Navigation).
- Finland—Finavia.
- France—DGAC (Government body).
- Germany—Deutsche Flugsicherung (German ATC).
- Greece—Hellenic Civil Aviation Authority (Hellenic ATC).
- Hong Kong—Civil Aviation Department.
- Ireland—IAA (Irish Aviation Authority).
- India—Airports Authority of India (AAI) (under Ministry of Civil Aviation, Government of India).
- Italy—ENAV (Italian ATC).
- Mexico—Servicios a la Navegación en el Espacio Aéreo Mexicano.
- Netherlands—LVNL (Dutch ATC).
- New Zealand—Airways Corporation (State Owned Enterprise).
- Norway—Avinor (State-owned private company).
- Pakistan—Civil Aviation Authority (State department).
- Singapore—CAAS (Civil Aviation Authority of Singapore).

- South Africa—Air Traffic and Navigation Services.
- Spain—AENA (Spanish ATC and Airports).
- Sweden—The LFV Group (Swedish ATC).
- Switzerland—Skyguide.
- United Kingdom—National Air Traffic Services (49% State Owned Public-Private Partnership).
- United States of America—Federal Aviation Administration (Government Body).
- Ukraine—Ukrainian State Air Traffic Service Enterprise (UkSATSE).

FLIGHT DATA RECORDER

The Flight Data Recorder (FDR) is a flight recorder used to record specific aircraft performance parameters. A separate device is the Cockpit Voice Recorder (CVR), although some recent types combine both in one unit. Popularly, though almost always falsely, known as the black box used for aircraft mishap analysis, the FDR is also used to study air safety issues, material degradation, and jet engine performance. These ICAO regulated "black box" devices are often used as an aid in investigating aircraft mishaps, and its recovery is second only in importance to the recovery of victims' bodies. The device's shroud is usually painted bright orange and generally located in the tail section of the aircraft.

History

The first prototype FDR was produced in 1957 by Dr. David Warren of the then Aeronautical Research Laboratories of Australia. In 1953 and 1954, a series of fatal mishaps on the De Havilland DH106 Comet prompted the grounding of the entire fleet pending an investigation. Dr. Warren, a chemist specializing in aircraft fuels, was involved in a professional committee discussing the possible causes. Since there had been neither witnesses or survivors, Dr. Warren conceived of a crash survivable method to record the flight crew's conversation, reasoning they would likely know the cause.

Despite his 1954 report entitled "A Device for Assisting Investigation into Aircraft Accidents" and a 1957 prototype FDR called "The ARL Flight Memory Unit", aviation authorities from around the world were largely uninterested. This changed in 1958 when Sir Robert Hardingham, the Secretary of the UK Air Registration Board, became interested. Dr. Warren was asked to create a pre-production model which culminated in the "Red Egg", the world's first commercial FDR, made by the British firm of S. Davall & Son. The "Red Egg" got its name from its shape and bright red colour. Incidentally, the term "Black Box" came from a meeting about the "Red Egg", when afterwards a journalist told Dr. Warren, *This is a wonderful black box.* The unit itself was based on a wire recorder fitted into a perspex box firmly screwed together. The first commercial airliner fitted with an FDR was the Hawker-Siddeley HS121 Trident.

Design

The design of today's FDR is largely governed by the European Organisation for Civil Aviation Equipment in its EUROCAE ED-112 (Minimum Operational Performance Specification for Crash Protected Airborne Recorder Systems). In the United States, the Federal Aviation Administration (FAA) regulates all aspects of U.S. aviation, and cites design requirements in their Technical Standard Order, based on the European ED-112 (as do the aviation authorities of many other countries).

Currently, EUROCAE specifies that a recorder must be able to withstand an acceleration of 3400 g (33 km/s^2) for 6.5 milliseconds. This is roughly equivalent to an impact velocity of 270 knots and a deceleration or crushing distance of 450 mm. Additionally, there are requirements for penetration resistance, static crush, high and low temperature fires, deep sea pressure, sea water immersion and fluid immersion.

Modern day FDRs are typically plugged into the aircraft's fly-by-wire main data bus. They record significant flight parameters, including the control and actuator positions, engine information and time of day. There are 88 parameters required as a minimum under current U.S. federal regulations (only 29 were required until 2002), but some systems monitor many more variables. Generally each parameter is recorded a few times per second, though some units store "bursts" of data at a much faster frequency if the data begins to change quickly. Most FDRs record 25 hours worth of data in a continuous loop.

This has also given rise to flight data monitoring programmes, whereby flights are analyzed for optimum fuel consumption and dangerous flight crew habits. The data from the FDR is transferred, *in-situ*, to a solid state recording device and then periodically analyzed with some of the same technology used for accident investigations.

FDRs are usually located in the rear of the aircraft, typically in the tail. In this position, the entire front of the aircraft acts as a "crush zone" to reduce the shock that reaches the recorder. Also, modern FDRs are typically double wrapped, in strong corrosion-resistant stainless steel or titanium, with high-temperature insulation inside.

Since the recorders can sometimes be crushed into unreadable pieces, or even never located in deep water, some modern units are self-ejecting (taking advantage of kinetic energy at impact to separate themselves from the aircraft) and also equipped with radio and sonar beacons (see emergency locator transmitter) to aid in their location. Future regulations may require such recorders on civil aircraft.

AIRLINE

An airline provides air transport services for passengers or freight. Airlines lease or own their aircraft with which to supply these services and may form partnerships or alliances with other airlines for reasons of mutual benefit.

Industry Overview

Airlines vary from those with a single airplane carrying mail or cargo, through full-service international airlines operating many hundreds of airplanes. Airline services can be categorized as being intercontinental, intracontinental, regional or domestic and may be operated as scheduled services or charters.

Patterns

- The pattern of ownership has gone from government owned or supported to independent, for-profit public companies. This occurs as regulators permit greater freedom and non-government ownership, in steps that are usually decades apart. This pattern is not seen for all airlines in all regions.
- The demand for air travel services depends on: business needs for cargo shipments, business passenger demand, leisure passenger demand, all influenced by economic activity.
- The overall trend of demand has been consistently increasing. In the 1950s and 1960s, annual growth rates of 15 per cent or more were common. Annual growth of 5-6 per cent persisted through the 1980s and 1990s. Growth

rates are not consistent in all regions, but countries with a de-regulated airline industry have more competition and greater pricing freedom. This results in lower fares and sometimes dramatic spurts in traffic growth. The U.S., Australia, Japan, Brazil, Mexico, India and other markets exhibit this trend.

- The industry is cyclical. Four or five years of poor performance preceed five or six years of improved performance. But profitability in the good years is generally low, in the range of 2-3 per cent net profit after interest and tax. In times of profit, airlines lease new generations of airplanes and upgrade services in response to higher demand. Since 1980, the industry has not earned back the cost of capital during the best of times. Conversely, in bad times losses can be dramatically worse.
- Warren Buffett once said that despite all the money that has been invested in all airlines, the net profit is less than zero. He believes it is one of the hardest businesses to manage.
- As in many mature industries, consolidation is a trend. Airline groupings may consist of limited bilateral partnerships, long-term, multi-faceted alliances between carriers, equity arrangements, mergers, or takeovers. Since governments often restrict ownership and merger between companies in different countries, most consolidation takes place within a country. In the U.S., over 200 airlines have merged, been taken over, or gone out of business since deregulation in 1978. Many international airline managers are lobbying their governments to permit greater consolidation to achieve higher economy and efficiency.

World's First Airline

DELAG, *Deutsche Luftschiffahrts-Aktiengesellschaft* (German: acronym for "German Airship Transport Corporation") was the world's first airline. It was founded on 16 November, 1909 with government assistance, and operated airships manufactured by Zeppelin Corporation. Its headquarters were in Frankfurt.

Early Development of Airlines in the U.S.

Tony Jannus conducted the United State's first scheduled commercial airline flight on 1 January, 1914 for the St. Petersburg-

Tampa Airboat Line. The 23 minute flight travelled between St. Petersburg, Florida and Tampa, Florida, passing some 50 feet above Tampa Bay in Jannus' Benoist biplane seaplane.

Following World War I, the United States found itself swamped with aviators. Many decided to take their war-surplus aircraft on barnstorming campaigns, performing acrobatic maneuvers to woo crowds. In 1918, the United States Postal Service won the financial backing of Congress to begin experimenting with air mail service, initially using Curtiss Jenny aircraft that had been procured by the United States Army for reconnaissance missions on the Western Front. Private operators were the first to fly the mail but due to numerous accidents the US Army was tasked with mail delivery. During the course of the Army's involvement they proved to be too unreliable and lost their air mail duties. By the mid-1920s, the Postal Service had developed its own air mail network, based on a transcontinental backbone between New York and San Francisco. To supplant this service, they offered twelve contracts for spur routes to independent bidders: the carriers that won these routes would, through time and mergers, evolve into Delta Air Lines, Braniff Airways, American Airlines, United Airlines (originally a division of Boeing), Trans World Airlines, Northwest Airlines, and Eastern Air Lines, to name a few.

Passenger service during the early 1920s was sporadic: most airlines at the time were focused on carrying bags of mail. In 1925, however, the Ford Motor Company bought out the Stout Aircraft Company and began construction of the all-metal Ford Trimotor, which became the first successful American airliner. With a 12-passenger capacity, the Trimotor made passenger service potentially profitable. Air service was seen as a supplement to rail service in the American transportation network.

At the same time, Juan Trippe began a crusade to create an air network that would link America to the world, and he achieved this goal through his airline, Pan American World Airways, with a fleet of flying boats that linked Los Angeles to Shanghai and Boston to London. Pan Am was the only U.S. airline to go international before the 1940s.

With the introduction of the Boeing 247 and Douglas DC-3 in the 1930s, the U.S. airline industry was generally profitable, even during the Great Depression. This trend continued until the beginning of World War II.

Early Development of Airlines in Europe

The first countries in Europe to embrace air transport were Finland, France, Germany and the Netherlands.

KLM, the oldest carrier still operating under its original name, was founded in 1919. The first flight transported two English passengers to Schiphol, Amsterdam from London in 1920. Like other major European airlines of the time, KLM's early growth depended heavily on the needs to service links with far-flung colonial possessions (Dutch Indies). It is only after the loss of the Dutch Empire that KLM found itself based at a small country with few potential passengers, depending heavily on transfer traffic, and was one of the first to introduce the hub-system to facilitate easy connections.

France began an air mail service to Morocco in 1919 that was bought out in 1927, renamed Aéropostale, and injected with capital to become a major international carrier. In 1933, Aéropostale went bankrupt, was nationalized and merged with several other airlines into what became Air France.

In Finland, the charter establishing Aero O/Y (now Finnair, one of the oldest still-operating airlines in the world) was signed in the city of Helsinki on 12 September, 1923. Junkers F 13 D-335 became the first aircraft of the company, when Aero took delivery of it on 14 March, 1924. The first flight was between Helsinki and Tallinn, capital of Estonia, and it took place on 20 March, 1924, one week later.

Germany's Lufthansa began in 1926. Lufthansa, unlike most other airlines at the time, became a major investor in airlines outside of Europe, founding Varig and Avianca. German airliners built by Junkers, Dornier, and Fokker were the most advanced in the world at the time. The peak of German air travel came in the mid-1930s, when Nazi propaganda ministers approved the start of commercial zeppelin service: the big airships were a symbol of industrial might, but the fact that they used flammable hydrogen gas raised safety concerns that culminated with the Hindenburg disaster of 1937. The reason they used hydrogen instead of the not-flammable helium gas was a United States military embargo on helium.

The United Kingdom's flag carrier during this period was Imperial Airways, which became BOAC (British Overseas Airlines Co.) in 1939. Imperial Airways used huge Handley-Page biplanes for routes between London, the Middle East, and India: images of

Imperial aircraft in the middle of the Rub'al Khali, being maintained by Bedouins, are among the most famous pictures from the heyday of the British Empire.

Development of Airlines Post-1945

As governments met to set the standards and scope for an emergent civil air industry toward the end of the war, it was no surprise that the U.S. took a position of maximum operating freedom. After all, U.S. airline companies were not devastated by the war, as European companies and the few Asian companies had been. This preference for "open skies" operating regimes continues, within limitations, to this day.

World War II, like World War I, brought new life to the airline industry. Many airlines in the Allied countries were flush from lease contracts to the military, and foresaw a future explosive demand for civil air transport, for both passengers and cargo. They were eager to invest in the newly emerging flagships of air travel such as the Boeing Stratocruiser, Lockheed Constellation, and Douglas DC-6. Most of these new aircraft were based on American bombers such as the B-29, which had spearheaded research into new technologies such as pressurization. Most offered increased efficiency from both added speed and greater payload.

In the 1950s, the De Havilland Comet, Boeing 707, Douglas DC-8, and Sud Aviation Caravelle became the first flagships of the Jet Age in the West, while the Soviet Union bloc countered with the Tupolev Tu-104 and Tupolev Tu-124 in the fleets of state-owned carriers such as Aeroflot and Interflug. The Vickers Viscount and Lockheed L-188 Electra inaugurated turboprop transport.

The next big boost for the airlines would come in the 1970s, when the Boeing 747, McDonnell Douglas DC-10, and Lockheed L-1011 inaugurated widebody ("jumbo jet") service, which is still the standard in international travel. The Tupolev Tu-144 and its Western counterpart, Concorde, made supersonic travel a reality. In 1972, Airbus began producing Europe's most commercially successful line of airliners to date. The added efficiencies for these aircraft were often not in speed, but in passenger capacity, payload, and range.

1978's U.S. airline industry deregulation lowered barriers for new airlines. In this period, new start-ups entered during downturns in the normal 8-10 year business cycle. At that time, they find aircraft, are financed, contract hangar and maintenance services, train new employees, and recruit laid off staff from other airlines.

As the business cycle returned to normalcy, major airlines dominated their routes through aggressive pricing and additional capacity offerings, often swamping new startups. Only America West Airlines (which has since merged with US Airways) remained a significant survivor from this new entrant era, as dozens, even hundreds, have gone under.

In many ways, the biggest winner in the deregulated environment was the air passenger. Indeed, the U.S. witnessed an explosive growth in demand for air travel, as many millions who had never or rarely flown before became regular fliers, even joining frequent flyer loyalty programmes and receiving free flights and other benefits from their flying. New services and higher frequencies meant that business fliers could fly to another city, do business, and return the same day, for almost any point in the country. Air travel's advantages put intercity bus lines under pressure, and most have withered away.

By the 1980s, almost half of the total flying in the world took place in the U.S., and today the domestic industry operates over 10,000 daily departures nationwide.

Toward the end of the century, a new style of low cost airline emerged, offering a no-frills product at a lower price. Southwest Airlines, JetBlue, AirTran Airways, and other low-cost carriers represent a serious challenge to today's legacy airlines, as do their low-cost counterparts in Europe, Canada, and Asia. Their commercial viability represents a serious competitive threat to the legacy carriers.

Thus, the last 50 years of the airline industry have varied from reasonably profitable, to devastatingly depressed. As the first major market to deregulate the industry in 1978, U.S. airlines have experienced more turbulence than almost any other country or region.

Regulatory Considerations

Government Regulation

Many countries have national airlines that are owned and operated by the Government. Even fully privatized airlines are subject to a great deal of government regulation for economic, political and safety concerns. Airline labour actions, for instance, are often halted by Government intervention in order to protect the free flow of people, communications, and goods between different regions without compromising safety.

The United States, Australia, and to a lesser extent Brazil, Mexico, the United Kingdom, and Japan have "deregulated" their airlines. In the past, these governments dictated airfares, route networks, and other operational requirements for each airline. Since deregulation, airlines have been largely free to negotiate their own operating arrangements with different airports, enter and exit routes easily, and to levy airfares and supply flights according to market demand.

The entry barriers for new airlines are lower in a deregulated market, and so the U.S. has seen hundreds of airlines start up (sometimes for only a brief operating period). This has produced far greater competition than before deregulation in most markets, and average fares tend to drop 20 per cent or more. The added competition, together with pricing freedom, means that new entrants often take market share with highly reduced rates that, to a limited degree, full service airlines must match. This is a major constraint on profitability for established carriers, which tend to have a higher cost base.

As a result, profitability in a deregulated market is uneven for most airlines. These forces have caused some major airlines to go out of business, in addition to most of the poorly established new entrants.

International Regulation

Groups such as the International Civil Aviation Organization establish worldwide standards for safety and other vital concerns. Most international air traffic is regulated by bilateral agreements between countries, which designate specific carriers to operate on specific routes. The model of such an agreement was the Bermuda Agreement between the US and UK following World War II, which designated airports to be used for transatlantic flights and gave each government the authority to nominate carriers to operate routes.

Bilateral agreements are based on the "freedoms of the air," a group of generalized traffic rights ranging from the freedom to overfly a country to the freedom to provide domestic flights within a country (a very rarely granted right known as cabotage). Most agreements permit airlines to fly from their home country to designated airports in the other country: some also extend the freedom to provide continuing service to a third country, or to another destination in the other country while carrying passengers from overseas.

In the 1990s, "open skies" agreements became more common, which take many of these regulatory powers from State Governments and open-up international routes to further competition. Open skies agreements have met some criticism, particularly within the European Union, whose airlines would be at a comparative disadvantage with the United States, because of cabotage restrictions.

Economic Considerations

Historically, air travel has survived largely through state support, whether in the form of equity or subsidies. The airline industry as a whole has made a cumulative loss during its 120-year history, once subsidies for aircraft development and airport construction are included in the cost.

The lack of profitability and continuing government subsidies are justified with the argument that positive externalities, such as higher growth due to global mobility, outweigh micro-economic losses. A historically high level of government intervention in the airline industry can be seen as part of a wider political consensus on strategic forms of transport, such as highways and railways, both of which are also publicly funded in most parts of the world. Profitability is likely to improve in future as privatization continues and more competitive low-cost carriers proliferate.

Although many countries continue to operate state-owned or parastatal airlines, many large airlines today are privately owned and are therefore governed by micro-economic principles in order to maximize shareholder profit.

Ticket Revenue

Airlines assign prices to their services in an attempt to maximize profitability. The pricing of airline tickets has become increasingly complicated over the years and is now largely determined by computerized yield management systems.

Most airlines use differentiated pricing, a form of price discrimination, in order to sell air services at varying prices simultaneously to different segments. Factors influencing the price include the days remaining until departure, the current booked load factor, the forecast of total demand by price point, competitive pricing in force, and variations by day of week of departure and by time of day. Carriers often accomplish this by dividing each cabin of the aircraft (first, business and economy) into a number of travel classes for pricing purposes.

A complicating factor is that of origin-destination control ("O&D control"). Someone purchasing a ticket from say, Melbourne to Sydney for $A200 is competing with someone else who wants to fly Melbourne to Los Angeles through Sydney on the same airplane, and who is willing to pay $A1400. Should the airline prefer the $A1400 passenger, or the $A200 passenger + a possible Sydney-Los Angeles passenger willing to pay $A1300? Airlines have to make hundreds of thousands of similar pricing decisions daily in their markets.

The advent of advanced computerized reservations systems in the late 1970s, most notably Sabre, allowed airlines to easily perform cost-benefit analyses on different pricing structures, leading to almost perfect price discrimination in some cases (that is, filling each seat on an aircraft at the highest price that can be charged without driving the consumer elsewhere). The intense nature of airfare pricing has led to the term "fare war" to describe efforts by airlines to undercut other airlines on competitive routes. Through computers, new airfares can be published quickly and efficiently to the airlines' sales channels. For this purpose the airlines use the Airline Tariff Publishing Company (ATPCO), who distribute latest fares for more than 500 airlines to Computer Reservation Systems across the world.

The extent of these pricing phenomena is strongest in "legacy" carriers. In contrast, low fare carriers usually offer pre-announced and simplified price structure, and often quote prices for each leg of a trip separately.

Computers also allow airlines to predict, with some accuracy, how many passengers will actually fly after making a reservation to fly. This allows airlines to overbook their flights enough to fill the aircraft while accounting for "no-shows," but not enough (in most cases) to force paying passengers off the aircraft for lack of seats. Since an average of 1/3 of all seats are flown empty, stimulative pricing for low demand flights coupled with overbooking on high demand flights can help reduce this figure.

Operating Costs

Full-service airlines have a high level of fixed and operating costs in order to establish and maintain air services: labour, fuel, airplanes, engines, spares and parts, IT services and networks, airport equipment, airport handling services, sales distribution, catering, training, aviation insurance and other costs. Thus all but a small percentage of the income from ticket sales is paid out to a wide variety of external providers or internal cost centers.

Moreover, the industry is structured so that airlines often act as tax collectors. Airline fuel is untaxed however due to a series of treaties existing between countries. Ticket prices include a number of fees, taxes, and surcharges they have little or no control over, and these are passed through to various providers. Airlines are also responsible for enforcing government regulations. If airlines carry passengers without proper documentation on an international flight, they are responsible for returning them back to the originating country.

Analysis of the 1992-96 period shows that every player in the air transport chain is far more profitable than the airlines, who collect and pass through fees and revenues to them from ticket sales. While airlines as a whole earned 6 per cent return on capital employed (2-3.5% less than the cost of capital), airports earned 10 per cent, catering companies 10-13 per cent, handling companies 11-14 per cent, aircraft lessors 15 per cent, aircraft manufacturers 16 per cent, and global distribution companies more than 30 per cent. (*Source*: Spinetta, 2000, quoted in Doganis, 2002)

In contrast, Southwest Airlines has been the most profitable of airline companies since 1970. Indeed, some sources have calculated Southwest to be the best performing stock over the period, outperforming Microsoft and many other high performing companies. The chief reasons for this are their product consistency and cost control.

The widespread entrance of a new breed of low cost airlines beginning at the turn of the century has accelerated the demand that full service carriers control costs. Many of these low cost companies emulate Southwest Airlines in various respects, and like Southwest, they are able to eke out a consistent profit throughout all phases of the business cycle.

As a result, a shakeout of airlines is occurring in the U.S. and elsewhere. United Airlines, US Airways (twice), Delta Air Lines, and Northwest Airlines have all declared Chapter 11 bankruptcy, and American has barely avoided doing so. Alitalia, Scandinavian Airlines System, SABENA, Japan Air System, Air Canada, Ansett Australia, and others have flirted with or declared bankruptcy since 2000, as low cost entrants enter their home markets as well. Some argue that it would be far better for the industry as a whole if a wave of actual closures were to reduce the number of "undead" airlines competing with healthy airlines while being artificially protected from creditors via bankruptcy law.

Where an airline has established an engineering base at an airport then there may be considerable economic advantages in using that same airport as a preferred focus (or "hub") for its scheduled flights.

Assets and Financing

Airline financing is quite complex, since airlines are highly leveraged operations. Not only must they purchase (or lease) new airline bodies and engines regularly, they must make major long-term fleet decisions with the goal of meeting the demands of their markets while producing a fleet that is relatively economical to operate and maintain. Compare Southwest Airlines and their reliance on a single airplane type (the Boeing 737 and derivatives), with the now defunct Eastern Air Lines which operated 17 different aircraft types, each with varying pilot, engine, maintenance, and support needs.

A second financial issue is that of hedging oil and fuel purchases, usually second only to labour in its relative cost to the company but with the current high fuel prices it has become biggest part of total airlines expenses. While hedging instruments can be expensive, they can easily pay for themselves many times over in periods of increasing fuel costs, such as in the 2000-05 period.

In view of the congestion apparent at many international airports, the ownership of slots at certain airports (the right to take-off or land an aircraft at a particular time of day or night) has become a significant tradable asset for many airlines. Clearly take-off slots at popular times of the day can be critical in attracting the more profitable business traveller to a given airline's flight and in establishing a competitive advantage against a competing airline. If a particular city has two or more airports, market forces will tend to attract the less profitable routes, or those on which competition is weakest, to the less congested airport, where slots are likely to be more available and therefore cheaper. Other factors, such as surface transport facilities and onward connections, will also affect the relative appeal of different airports and some long distance flights may need to operate from the one with the longest runway.

Airline Partnerships

Code sharing is the most common type of airline partnership; it involves one airline selling tickets for another airline's flights under its own airline code. An early example of this was Japan Airlines' code sharing partnership with Aeroflot in the 1960s on flights from Tokyo to Moscow: Aeroflot operated the flights using Aeroflot aircraft, but JAL sold tickets for the flights as if they were JAL flights. This practice allows airlines to expand their operations, at least on paper, into parts of the world where they cannot afford to establish bases or purchase aircraft.

Since airline reservation requests are often made by city-pair (such as "show me flights from Chicago to Düsseldorf"), an airline who is able to code share with another airline for a variety of routes might be able to be listed as indeed offering a Chicago-Düsseldorf flight. The passenger is advised however, that Airline 1 operates the flight from say Chicago to Amsterdam, and Airline 2 operates the continuing flight (on a different airplane, sometimes from another terminal) to Düsseldorf. Thus, the primary rationale for code sharing is to expand one's service offerings in city-pair terms so as to increase sales.

Virtually all international airlines practice code sharing.

A more recent development is the airline alliance, which became prevalent in the 1990s. These alliances can act as virtual mergers to get around government restrictions. Groups of airlines such as the Star Alliance, Oneworld, and SkyTeam coordinate their passenger service programmes (such as lounges and frequent flyer programmes), offer special interline tickets, and often engage in extensive codesharing (sometimes systemwide). These are increasingly integrated business combinations—sometimes including cross-equity arrangements—in which products, service standards, schedules, and airport facilities are standardized and combined for higher efficiency. One of the first airlines to start an alliance with another airline was KLM, who partnered with Northwest Airlines. Both airlines later entered the SkyTeam alliance after the fusion of KLM and Air France in 2004.

Often the companies combine IT operations, buy fuel, or purchase airplanes as a bloc in order to achieve higher bargaining power. However, the alliances have been most successful at purchasing invisible supplies and services, such as fuel. Airlines usually prefer to purchase items visible to their passengers to differentiate themselves from local competitors. If an airline's main domestic competitor flies Boeing airliners, then the airline may prefer to use Airbus aircraft regardless of what the rest of the alliance chooses.

Customs and Conventions

Each operator of a scheduled or charter flight uses a distinct airline call sign when communicating with airports or air traffic control centers. Most of these call-signs are derived from the airline's trade name, but for reasons of history, marketing, or the need to reduce ambiguity in spoken English (so that pilots do not mistakenly make navigational decisions based on instructions issued to a different aircraft), some airlines and air forces use call-signs less obviously connected with their trading name. For example, British Airways uses a *Speedbird* call-sign, named after the logo of its predecessor, BOAC.

Airline Personnel

The various types of airline personnel include:

- Flight crews, responsible for the operation of the aircraft. Flight crew members include:
  - Pilots (Captain and First Officer: some older aircraft also require a Flight Engineer/Second Officer and/or Navigators);
  - Flight Attendants (led by a purser on larger aircraft); and
  - In-flight Security Personnel on some airlines (most notably El Al).
- Ground Crews, responsible for operations at airports. Ground crew members include:
  - Airframe and Powerplant technicians. Often the ratings are termed as 'A' and 'P' technicians;
  - Avionics technicians/engineers
  - Flight Dispatchers
  - Baggage Handlers

- Rampers
- Gate Agents
- Ticket Agents
- Passenger Service Agents (such as airline lounge employees).

■ Reservations Agents, usually (but not always) at facilities outside the airport.

Most airlines follow a corporate structure where each broad area of operations (such as maintenance, flight operations, and passenger service) is supervised by a vice-president. Larger airlines often appoint vice-presidents to oversee each of the airline's hubs as well. Airlines also tend to employ considerable numbers of lawyers to deal with regulatory procedures and other administrative tasks.

UNITED AIRLINES FLIGHT 232: A CASE STUDY

United Airlines Flight 232 was a scheduled flight operated by United Airlines between Denver and Philadelphia *via* Chicago. On 19 July, 1989, the Douglas DC-10 (Registration N1819U) operating this flight suffered an uncontained failure of its number 2 engine (mounted in the tail), which destroyed all three of the aircraft's hydraulic systems. With no controls working except the power levers for the two remaining engines, it broke up during an emergency landing on the runway at Sioux City, Iowa killing 110 of its 285 passengers and one of the 11 crew members. It is one of the best-recognized air disasters in aviation history, largely due to the presence of television crews near the airfield as the plane landed and broke up in a large fireball.

Owing to the skill of the crew and a DC-10 instructor pilot, 184 passengers and 10 crew members survived. The crash is considered to be a textbook example of successful Crew Resource Management, due to the effective use of all the resources available aboard the plane for help during the emergency.

Chronology of the Flight

The flight took off at 14:09 (CDT) from Stapleton International Airport, Denver, Colorado and was due to fly to Philadelphia International Airport in Philadelphia, Pennsylvania *via* O'Hare International Airport in Chicago, Illinois. At 15:16, while the plane was in a shallow right turn at 37,000 feet, the N1 stage fan disk of its tail-mounted General Electric CF6-6 engine broke in two. The fan cowling was blown off and pieces of the engine penetrated the aircraft tail section in numerous places, including both horizontal stabilizers. Pieces of shrapnel which went through the right horizontal stabilizer severed the lines of all three hydraulic systems, allowing the fluid to drain away.

Captain Alfred C. Haynes and his flight crew (First Officer William Records, who was flying, and Second Officer Dudley Dvorak, flight engineer) felt a jolt going through the aircraft, and warning lights showed that the autopilot had disengaged, and the tail-mounted number two engine was malfunctioning. The co-pilot noticed that the airliner was off course, and moved his control column to correct this, but the plane did not respond. The flight crew discovered that the pressure gauges for each of the three hydraulic systems was registering zero, and they realized that the initial failure had left all control surfaces immovable.

As the failure of all three hydraulic systems had been considered extremely unlikely by the aircraft engineers, there was no backup means of safely controlling the aircraft. It had a continual tendency to turn right, and was difficult to maintain on a stable course. The plane began to slowly oscillate vertically in a phugoid cycle, which is characteristic of accidents where hydraulic power was lost. With each iteration of the cycle the aircraft lost approximately 1500 feet of altitude. Dennis E. Fitch, a DC-10 flight instructor, was deadheading as a passenger on the plane and offered his assistance. After entering the cockpit, Fitch discovered that the only available method of controlling the aircraft was through adjusting the throttles of the remaining two engines; running one engine faster than the other to turn the plane, and accelerating or decelerating in order to gain or lose altitude. Using this method, it was possible to mitigate the phugoid cycle and make rough steering adjustments. At one point Fitch manually lowered the landing gear in flight, hoping that this would force trapped hydraulic fluid back into the lines allowing some movement of control surfaces. Although the gears lowered successfully, there was no improvement in control response as all the fluid had been lost through the punctured lines.

Air Traffic Control (ATC) was contacted and an emergency landing at nearby Sioux Gateway Airport was organized.

Haynes kept his sense of humor during the emergency, as recorded on the plane's CVR:

Fitch: *I'll tell you what, we'll have a beer when this is all done.* Haynes: *Well I don't drink, but I'll sure as hell have one.*

and later:

Sioux City Approach: *United Two Thirty-Two Heavy, the wind's currently three six zero at one one; three sixty at eleven. You're cleared to land on any runway.* Haynes: *[laughter] Roger. [laughter] You want to be particular and make it a runway, huh?*

A more serious remark often quoted from Haynes was made when ATC asked the crew to make a left turn to keep them clear of the city:

Whatever you do, keep us away from the city.

Landing was originally planned on the 9,000 feet (2743 m) Runway 31. The difficulties in controlling the aircraft made lining up almost impossible. While dumping excess fuel, the plane executed a series of mostly right-hand turns (it was easier to turn the plane in this direction) with the intention of coming out at the end lined up with runway 31. When they came out they were instead left with an approach on the shorter Runway 22 of 6,600 feet (2012 m), with little capacity to maneuver.

Fire trucks had been placed on runway 22, anticipating a landing on runway 31, and there was a scramble as the trucks rushed out of the way. Fortunately, all the vehicles parked there got out of the way before the plane touched down.

Fitch continued to control the aircraft's descent by adjusting engine thrust. With the loss of all hydraulics, the crew were unable to control airspeed independent from sink rate. On final descent, the aircraft was going 240 knots and sinking at 1850 feet per minute, while a safe landing would require 140 knots and 300 feet per minute. The aircraft began to sink faster while on final approach and veer to the right. The tip of the right wing hit the runway first, spilling fuel which ignited immediately. The tail section broke off from the force of the impact and the rest of the aircraft bounced several times, shedding the landing gear and engine nacelles and breaking the fuselage into several main pieces. On the final impact the right wing was sheared off and the main part of the aircraft skidded sideways, rolled over on to its back, and slid to a stop upside down on the right side of runway 22. A misconception that the aircraft "cartwheeled" wingtip-over-wingtip arose from the poor quality of the video and news reporters' initial interpretation of it.

111 people died in the crash. Most were killed by injuries sustained in the multiple impacts, but many in the middle fuselage section directly above the fuel tanks died from smoke inhalation in the post-crash fire, which burned for longer than it might due to delays in the firefighting efforts. 185 people survived the accident, the majority of them from the second-class seating ahead of the wings. Many passengers were able to walk out through the ruptures to the structure, and in many cases got lost in the high field of corn their part of the plane had ended up in. As with the Eastern Air Lines Flight 401 crash of a similarly-sized Lockheed L-1011 in 1972, the inherent crashworthiness of newer wide-bodied air transports played a part in the relatively high survival rate, as well as the shallow rate of descent and relatively low speed.

Causes

Investigation assigned the origin of the fracture of the fan disk to a failure of United Airlines maintenance processes to detect an existing fatigue crack. Post-crash analysis of the crack surfaces showed the presence of the penetrating fluorescent dye used to detect cracks during maintenance, indicating that the crack was present and should have been detected at a prior inspection. The detection failure arose from poor attention to human factors in United Airlines' specification of maintenance processes.

The crack was due to a metal 'inclusion' in the disk and was traced back to the metal processing plant from which the engine part was sourced. It turned out that there was a defect in elimination of gaseous anomalies during the purifying of the (molten) titanium disk ingot. These anomalies, if left unchecked, would contribute to a systematic decrease of the metal's strength over time and, ultimately, its failure. Newer batches used a 'triple vacuum' process to eliminate these impurities.

The subsequent investigation and airworthiness directive also revealed several other fan disks already in service from the same batch of ingots which had started to exhibit the initial cracking symptoms of part failure.

Some portions of the plane that broke away when the fan disk failed were later found in farm fields along the flight path.

Lessons Learned

In subsequent reconstructions of the accident in flight simulators, it was deemed that training for such an event involved too many factors to be practical. While some level of control was possible, no precision could be achieved and a landing under these conditions was stated to be "a highly random event."

Because this type of aircraft control is difficult for humans to achieve, some researchers have attempted to integrate this control ability into the computers of fly-by-wire aircraft. Early attempts to add the ability to real planes were not very successful, the software having been based on experiments conducted in flight simulators where jet engines are usually modeled as "perfect" devices with exactly the same thrust on each engine. Later, programming was updated to compensate for the problem, and planes have been successfully flown with this software installed. However, it remains a rarity on commercial aircraft.

The odds against all three hydraulic systems failing simultaneously had previously been calculated as high as a billion to one, but a similar failure had occurred just four years previously when Japan Airlines flight 123 suffered a structural failure that left it without any hydraulic controls.

Newer aircraft designs such as the MD-11 have incorporated hydraulic fuses to isolate a punctured section and prevent a total loss of hydraulic fluid. This was also partially implemented on DC-10 models after the accident.

Of the four children deemed too young to have seats of their own (so-called 'lap children'), one died from smoke inhalation. Despite the survival rate for the lap children being much higher than that of the passengers (75% *versus* 62.5%), the NTSB added a safety recommendation to the FAA on its "List of Most Wanted Safety Improvements," the response to which was flagged on NTSB's website as an "acceptable response, progressing slowly." It also sparked a campaign, led by United Flight 232's senior flight attendant, Jan Brown Lohr, for all children to have seats on aircraft. Lohr continues to lobby in Washington D.C. and many flight attendants from various airlines support her cause.

"Three Contributing Factors"

Of the 296 people aboard, 111 were killed as a result of the crash. 185 survived the incident. Captain Haynes later told of "three contributing factors" regarding the actual time of day that allowed for a better chance of survival:

1. The incident occurred during daylight hours;

2. The incident occurred as a shift change was occurring at both a regional trauma center and a regional burn center in Sioux City, allowing for more medical personnel to treat the injured; and
3. The incident occurred when the Iowa Air National Guard was on duty at Sioux Gateway Airport, allowing for 285 trained personnel to assist with triage and evacuation of the wounded.

1994 Memorial

A statue in part of Sioux City's riverfront development depicts Lt. Colonel Dennis Nielsen carrying three-year-old survivor Spencer Bailey to safety. The memorial commemorates the rescue efforts shown by the Sioux City community after the crash. It features contemplative areas and a tree-lined approach with plaques narrating the crash.

Notable Survivors

- *Jerry Schemmel*: Radio announcer for the Denver Nuggets, an NBA basketball team. Also wrote a book about United Airlines Flight 232 titled *Chosen to Live*, and was credited for saving the life of a child in the crash.
- *Michael Matz*: Trainer of the 2006 Kentucky Derby favourite and winner, Barbaro. Also credited with saving the lives of four children in the crash, three of whom were in the same family.
- *Dennis E. Fitch*: A safety consultant to NASA as a member of the Aerospace Safety Advisory Panel and has also been inducted into the Aviation Hall of Fame at the Smithsonian National Air and Space Museum. Also a DC-10 pilot and instructor, and helped Captain Al Haynes fly United Airlines Flight 232.
- *Joseph Trombello*: Wrote a book about United Airlines Flight 232 titled *Miracle in the Cornfield.*
- *Jan Brown Lohr*: United 232's Senior Flight Attendant. She was forced by regulation to ask parents with "lap babies" (children without seats) aboard flight 232 to place their children on the cabin floor during the flight's final moments before impact. Upon impact one of four children was killed. The deceased child's mother came to her at the crash site and stated, "You told me to put my baby on the

floor and now he's gone." Since then, Lohr has tirelessly lobbied in Washington D.C. to promote the safety of children on all civilian aircraft and airlines, asking that federal regulations require all children to have a seatbelt on every flight.

In Popular Culture

The incident was made the subject of the 1992 television movie, *Crash Landing: The Rescue of Flight 232* (also known as *A Thousand Heroes*), and was also featured in an episode of *Seconds from Disaster* on the National Geographic Channel and *MSNBC Investigates* on the MSNBC news channel. It was also one of the inspirations for Peter Weir's 1993 film version of *Fearless*, adapted from a novel by Rafael Yglesias.

2

Aviation Management: Organizations, Regulations and Policies

AVIATION MANAGEMENT: AN INTRODUCTION

Aviation Managers have diverse responsibilities depending on which career track they choose. There are operational and support-based aviation management opportunities with large commercial carriers, at airports, and with smaller fixed-base operators. There are also careers in business aviation and aircraft manufacturing. Most aviation managers have the responsibility for overseeing the day-to-day operations of their department. Utilizing business principles to oversee staff, make budget decisions, and even market products and services can often be part of the aviation manager's job.

Aviation Maintenance Technician

Aviation Maintenance Technicians keep aircraft safe and in flying condition. They are responsible for having a detailed knowledge of all the systems on the aircraft. Aviation maintenance technicians may be sitting at computers translating technical data, replacing a reciprocating or turbine engine, or making a repair that maintains the structural integrity of a fuselage. Rapidly changing aircraft technology means that they are constantly updating their skills and knowledge to maintain aircraft in safe flying condition.

Global Air Traffic Management

Global Air Traffic Management (GATM) is a concept for satellite-based communication, navigation, surveillance and air traffic management. The Federal Aviation Administration and the International Civil Aviation Organization, a special agency of the United Nations, established GATM standards in order to keep air travel safe and effective in increasingly crowded worldwide air space.

Efforts are made all over the world to implement the new technologies that will allow GATM to efficiently support Air Traffic Controller.

Airservices Australia ADS-B initiative is one of the major implementation programme in this field. This initiative will facilitate the certification of this new technology allowing further implementation.

INTERNATIONAL CIVIL AVIATION ORGANIZATION

The International Civil Aviation Organization (ICAO), an agency of the United Nations, codifies the principles and techniques of international air navigation and fosters the planning and development of international air transport to ensure safe and orderly growth.

The ICAO Council adopts standards and recommended practices concerning air navigation, prevention of unlawful interference, and facilitation of border-crossing procedures for international civil aviation.

In addition, the ICAO defines the protocols for air accident investigation followed by transport safety authorities in countries signatory to the Convention on International Civil Aviation, commonly known as the Chicago Convention. See NTSB, TSB, AAIB, BFU and BEA.

Its headquarters are located in the *Quartier International* of Montreal, Canada.

The ICAO should not be confused with the International Air Transport Association (IATA), a trade organization for airlines also headquartered in Montreal.

ICAO Codes

Both ICAO and IATA have their own airport and airline code systems. ICAO uses 4-letter airport codes, and 3-letter airline codes. IATA is scheduled to eventually switch its codes to the ICAO standard. In the United States, the ICAO codes are usually the same as the IATA code, with a prefix of "K"—LAX is KLAX. In the rest of the world the codes are unrelated, as the IATA code is phonic and the ICAO code is location-based; for example, **Charles de Gaulle** ICAO:LFPG, IATA:CDG.

ICAO is also responsible for issuing alphanumeric aircraft type codes that contain 3 or 4 characters. These codes provide the identification that is typically used in flight plans. An example of this is the Boeing 747 that would use (depending on the variant) **B741**, **B742**, **B743**, etc.

ICAO provides telephony designators to aircraft operators worldwide. These consist of the three letter airline identifer and a one or two word designator. They are usually, but not always, similar to the aircraft operator name. Thus, the identifer for Aer Lingus is **EIN** and the designator is **Shamrock** while International is **JAL** and **Japan Air**. So a flight by Aer Lingus numbered 111 would be written as "EIN111" and pronounced "Shamrock One-One-One" on the radio, while a similarly numbered Japan Airlines flight would be written as "JAL111" and "Japan Air One-One-One."

ICAO maintains the standards for aircraft registration ("tail numbers"), including the alphanumeric codes that identify the country of registration.

ICAO Statute

The 8th edition of the Convention on International Civil Aviation includes modifications from 1948 upto year 2000. The ICAO refers to its current edition of the convention the *statute*, and designates it as ICAO Doc 7300/8.

The convention had 18 annexes:

- Annex 1—Personnel Licensing
- Annex 2—Rules of the Air
- Annex 3—Meteorological Service for International Air Navigation
- Annex 4—Aeronautical Charts
- Annex 5—Units of Measurement to be Used in Air and Ground Operations
- Annex 6—Operation of Aircraft—Aeroplanes
- Annex 7—Aircraft Nationality and Registration Marks
- Annex 8—Airworthiness of Aircraft
- Annex 9—Facilitation
- Annex 10—Aeronautical Telecommunications
- Annex 11—Air Traffic Services
- Annex 12—Search and Rescue
- Annex 13—Aircraft Accident and Incident Investigation
- Annex 14—Aerodromes—Aerodrome Design and Operations, Heliports

- Annex 15— Aeronautical Information Services
- Annex 16— Environmental Protection—Aircraft Noise
- Annex 17— Security: Safeguarding International Civil Aviation Against Acts of Unlawful Interference
- Annex 18— The Safe Transport of Dangerous Goods by Air.

ICAO Standards

The ICAO also standardizes certain functions for use in the airline industry, such as the Aeronautical Message Handling System AMHS; this probably makes it a standards organization.

Regions and Regional Offices

The ICAO has seven regional offices serving nine regions:

1. Asia and Pacific, Bangkok, Thailand.
2. Middle East, Cairo, Egypt.
3. Western and Central Africa, Dakar, Senegal.
4. South America, Lima, Peru.
5. North America, Central America and Caribbean, Mexico City, Mexico.
6. Eastern and Southern Africa, Nairobi, Kenya.
7. Europe and North Atlantic, Paris, France.

ICAO Leadership

List of Secretaries General

- Albert Roper (1944-1951).
- Carl Ljungberg (1952-59).
- Ronald MacAllister Macdonnell (1959-1964).
- Bernardus Tielman Twigt (1964-1970).
- Assad Kotaite (1970-76).
- Yves Lambert (1976-1988).
- Shivinder Singh Sidhu (1988-1991).
- Philippe Rochat (1991-97).
- Renato Claudio Costa Pereira (1997-2003).
- Taieb Cherif (2003-Present).

List of Council Presidents

- Edward Pearson Warner (1947-1957).
- Walter Binaghi (1957-1976).
- Assad Kotaite (1976-2006).
- Roberto Kobeh Gonzalez (2006-Present).

President of the Council

Opening address by the President of the Council of the International Civil Aviation Organization (ICAO), Mr. Roberto Kobeh González to the Seventh Meeting of the Committee on Aviation Environmental Protection (CAEP). Montréal, 5 February, 2007.

Secretary General

Keynote address by the Secretary-General of the International Civil Aviation Organization (ICAO), Dr. Taïeb Chérif, to the International Conference on Aviation. Bangalore, India, 8 and 9 February, 2007.

News Releases

Participants of ICAO Environmental Meeting agree on guidance for aviation emissions trading (16 February, 2007).

ICAO Update

A recent field mission by an ICAO Air Navigation Commission team has highlighted the need for increased ICAO involvement in coordinating and managing safety efforts throughout the African region.

Worldwide Symposium on Performance of the Air Navigation System

A Worldwide Symposium on Performance of the Air Navigation System will be held at the ICAO Headquarters on 26-30 March, 2007.

ICAO/AU/ATAG/WB Developement Forum

ICAO, the African Union, the Air Transport Action Group and the World Bank will hold a joint development forum on Maximizing Civil Aviation's Economic Contribution in African Skies (Challenges and Potentials) at the headquarters of the African Union in Addis Ababa, Ethiopia from 23 to 25 April, 2007.

Second Aviation Language Symposium (IALS/2)

The Second ICAO Aviation Language Symposium (IALS/2) will be held at the ICAO Headquarters on 7-9 May, 2007.

ICAO Colloquium on Aviation Emissions with Exhibition

ICAO invites to its second Environmental Colloquium at ICAO Headquarters in Montreal from 14 to 16 May, 2007. It will provide a forum on aviation emissions including key developments emanating from CAEP/7.

ICAO Journal

No. 1, January/February, 2007—The issue of runway incursions has been identified as a serious threat to aviation safety that many States are working hard to resolve. Recent initiatives to enhance runway safety include the implementation of a European action plan formulated by a group of experts familiar with the day-to-day problems on the job.

ICAO eSHOP

ICAO eSHOP is a Web-based information service. This service offers ICAO customers online access to various sets of ICAO documentation for an annual fee. Subscriptions give access to the full texts of International Conventions and Protocols, Annexes to the Convention on International Civil Aviation, documents pertaining to Air Traffic Management and the Annual Reports of the Council.

About ICAO

- Annual Reports of the Council (excerpts).
- Chicago Convention.
- Contracting States.
- History of international civil aviation.
- How it works.
- ICAO Council Members.
- ICAO emblem and its history (pdf).
- Making a Standard.
- Memorandum on ICAO (pdf).
- Premises.
- Office of the Secretary General.
- President of the Council.
- Programme Budget and Financial Statements.
- Strategic Objectives.
- Structure of ICAO Secretariat (pdf): en | fr | es | ru | ar | zh.

Key Objectives

- Flight Safety Information Exchange (FSIX).
- Aviation Security.
- Environment.
- Safety Oversight Audit.
- Strategic Objectives.
- Universal Security Audit Programme.

AVIATION.ORG: A CASE STUDY

History

Founded in 1995, Aviation Safety Connection, Inc. undertook the task of fostering improved flight safety by (1) promoting a better understanding of operational judgement and decision-making, and (2) presenting programmes and methods that can have a positive effect in achieving safety improvements.

The motivation for undertaking this mission, described more fully in the Director's Letter, arises from the observation that many aircraft accidents are repetitive in nature, the result of errors in human judgement. Based on these similarities, it is apparent that the underlying issues are not well-understood by those who have the ultimate responsibility for accident prevention, aircraft pilots.

Aviation.Org was launched by Aviation Safety Connection, Inc. in early 1996 to provide an interactive forum for pilots to analyze, discuss and reflect upon leadership and management in the cockpit. Although the readership had expanded substantially, numerically and geographically, the effort had to be curtailed for a short period of time. As of January, 2003, the website has been renewed and reinvigorated with greater emphasis and improved structure.

Join Us

Join Us to help make this endeavour a worthy and valuable undertaking. Registration is required to subscribe—however, there is no associated fee.

Sponsorship

Aviation Safety Connection, Inc. is a non-profit charitable and educational organization exempted from federal income tax under Section 501 (c)(3) of the Internal Revenue Code. The organization depends on the financial support of companies, associations and individuals to fulfil its objective of improving flight safety.

Data Bank

NTSB, NASA & FAA Accident, Incident and Miscellaneous Databases.

NTSB Accident Reports

The National Transportation Safety Board (NTSB) investigates all aircraft accidents and issues a formal report determining probable cause at the completion of each investigation. These reports are public documents. In addition to its fully documented reports, the NTSB maintains a data base of report summaries, called either "briefs" or "synopses." Dating from 1983, over 33,000 reports have been compiled and these synopses are posted to the NTSB website. This site has a search (query) capability, or a report can be found by the date and location of the accident.

ASRS Incident Reports

The FAA has contracted with the National Aeronautics and Space Administration (NASA) to collect and analyze aviation incident reports that are submitted voluntarily and anonymously. Under this Aviation Safety and Reporting System (ASRS), NASA maintains a data base of over 70,000 ASRS incident reports dating from 1986.

Access is available to selected NASA ASRS incident files. Reports are grouped by approximately 20 frequently requested topics with each group numbering 50 recent reports. Reports are updated quarterly.

FAA National Aviation Safety Data Analysis Center (NASDAC)

NASDAC maintains searchable databases, including NTSB and ASRS reports. Other databases include Air Registry (AR), Bureau of Transportation Statistics (BTS), FAA Accident/Incident Data System (AIDS), Near Midair Collision System (NMACS), NTSB Safety Recommendations, and World Aircraft Accident Summary (WAAS).

Taxi Clearance

Access to Hangar 13.

Cleared Direct

Couldn't be easier. The only directive is to fill in the Subscription Form and you're on your way. You'll be asked for an ID (your e-mail address) and a password of your choosing. Once you become a registered subscriber you will have Login access to Hangar 13.

Hangar 13 Activities

Hangar 13 is the "home page" of the subscriber section. Located within the hangar is the Ready Room with its document Library, and the Pilot Lounge.

Library

The recommended first stop is the Library located in the Ready Room. Here there is the Cockpit Leadership bookshelf containing reference documents on topics of current interest, including those applicable to ongoing Ready Room discussions.

The bookshelf has a series of Cockpit Leadership articles in a pilot's handbook format. These articles were written to serve as checklists and provide guidance for Ready Room discussions and commentary.

Ready Room

The Ready Room provides for active dialogue, briefings and debriefings. These interactive discussions are moderated and cover pilot performance issues on selected topics, including those suggested by subscribers. Topics are usually, but not necessarily, based on incident or accident reports.

In addition to the opportunity to participate in current discussions, subscribers will periodically receive a brief e-mail message, Cockpit Concepts, intended to keep ours focus on human performance issues of interest.

Pilot Lounge

The Pilot Lounge is a more informal setting. It has a bulletin board for dialogue on topics initiated by subscribers. It's requested that these topics and the comments posted be related to pilot performance issues. We want to stay our course, and numerous websites are available for discussion of other aviation issues.

Archives

This Library annex houses additional reference documents of general appeal that are related to cockpit issues.

AEROPOL AVIATION SERVICES CORPORATION

Aeropol Aviation Services Corporation was established at the beginning of 1998 to provide services for the requirements of aviation companies. Our focus is on hydraulic, pneumatic, electro-mechanical and fuel components. We are committed to delivering quality products and outstanding service by utilizing our capabilities through teamwork and focusing on customer satisfaction. Our technical team knowledge exceeds 20 years of aviation maintenance, which is added value to each part we service. Our facility is fully equipped and capable to service a wide range of commercial and military components. We pride ourselves on being able to provide fast turn around time, and on being able to offer a reliable and competitively priced service. The quality of work and the satisfaction of our customers remain our number one priority. We are large enough to support your MRO needs, yet small enough to attend to the smallest detail. Aeropol Aviation Services Corporation offers wide range capability to repair and overhaul hydraulic, pneumatic, fuel and electro-mechanical components such as: valves, actuators, pumps and gearboxes. Our standard turn around time is 7-10 days. AOG requests can be accommodated in 2-3 days (excluding shipping). Inventory is available for exchange or outright sale. We deal with major parts manufacturers as well as small suppliers to offer competitive prices and best meet our obligations. All spare parts used in component repair have acceptable airworthiness traceability. Since we provide our services to the international community (including Europe, East Asia and The Middle East), we comply with international standards and regulations to meet our customers' airworthiness requirements.

AVIATION MANAGEMENT CONSULTING GROUP

We provide a full range of aviation management consulting services to *fixed base operators, specialized aviation businesses, airport owners and operators, governmental entities and professional organizations.*

FBO, SASO, and general aviation-related services include:

- Valuation
  - Business

  - Stock
  - Asset.
- Appraisal
  - Leasehold Interest
  - Leased Fee.
- Acquisition.
- Divestiture.
- Operational and Financial Assessment.
- RFP Response Development (Proposal and Presentation)
- Strategic Business Planning.
- Market Assessment and Feasibility Study.
- Marketing and Business Development.
- Site Planning.
- Facility Programming.
- Litigation Support and Expert Testimony.

These Services are provided with the objective of:

- Enhancing revenues.
- Reducing and/or eliminating costs.
- Improving the quality and delivery of products and services.
- Maximizing efficiency, productivity, and safety.
- Increasing the value of a business.

Airport related services conclude:

- Primary Guiding Documents.
  - Minimum Standards.
  - Rules and Regulations.
  - Lease/Rates and Charges Policy.
  - Development Guidelines.
  - Lease, Use and Operating Agreements.
- Appraisal
  - Leasehold Interest
  - Leased Fee.
- Valuation
  - Business
  - Stock
  - Asset.
- RFP Development and Response Evaluation.
- Lease Rate Analysis (Rates and Charges).
- Market Assessment and Feasibility Study.

- Operational and Financial Assessment.
- Strategic Business Planning.
- Site Planning.
- Facility Programming.
- Litigation Support and Expert Testimony.

These Services are provided with the goal of:

- Improving tenant relations.
- Enhancing consumer products and services.
- Maximizing efficiency, productivity, and safety.
- Increasing operating cash flow.
- Attracting and retaining quality general aviation service providers (and consumers).

Management: Scope of Services

Valuation

This process is used to estimate the market value of an FBO or specialized aviation business which is predicated upon:

1. The ability of management to generate positive earnings utilizing the assets of the company.
2. The terms and conditions of the company's lease, use, or operating agreement.
3. Market conditions.
4. Industry sales transactions.

To estimate the value of an FBO or specialized aviation business, the **Aviation Management Consulting Group** conducts a comprehensive review of the operational and financial performance of the company. This includes a retrospective analysis of the financial results achieved by the company and a prospective look at the future. This review provides the opportunity to identify and evaluate each element of the company's income stream (and more specifically, to fully understand how the company's assets are being used to generate earnings).

By way of comparison to verified industry sales transactions, an appropriate earnings multiple is established. The multiple is then adjusted to account for various tangible (quantifiable) and intangible (qualitative) variables which have a direct impact on value. Once adjusted, the multiple is applied to the company's earnings to derive a market value estimate.

The market value estimate is frequently utilized to facilitate the sale, acquisition, or financing of an FBO or specialized aviation business.

Stock

We estimate the market value of preferred and common stock including majority and minority interest (s). Market value is based upon the expectation of the company's future profitability and dividends per share, the company's riskiness and financial condition, and the state of the money market.

Assets

We estimate the market value of vehicles, equipment, aircraft, and inventory on a liquidated, wholesale, or retail basis.

Appraisal

This process is used to determine the value of the interest held by a lessee (leasehold interest) or lesser (leased fee) in airport-based land and/or improvements.

Leasehold Interest

Leasehold interest is defined as "the interest held by the tenant (lessee) through a lease conveying the rights of use and occupancy for a stated term under certain conditions."

Leased Fee

Leased fee is defined as "the interest in a property that remains with the owner (lesser) after the right to use and occupy the property has been conveyed to another." This right is usually granted by a lease for a specific period of time.

An appraisal is typically performed when a value is desired or required (i.e., to facilitate the sale or acquisition of airport-based land and/or improvements or if a value is required by the lease, use, or operating agreement to calculate a rental rate).

From a theoretical standpoint, there are three fundamental approaches to value: cost, sales comparison, and income. Each of these approaches are discussed, as follows:

Cost Approach: This approach considers the current cost of replacing facilities and site improvements *less* depreciation *plus* the market value of the land (assumed vacant). The *cost approach* is most effective in valuing relatively new developments. Since improvements are typically encumbered by a land lease, leasehold value needs to be analyzed. In addition, when appraising airport-based properties, surrounding (off-airport) land values are considered; however, such values are not heavily weighted since there are significant differences between off-airport and on-airport properties.

Sales Comparison: This approach involves direct comparisons of similar properties that have sold in the same or similar markets. The data from comparable sales is analyzed and adjusted for differences that are considered significant. The adjusted sales are then weighted to provide an indication of value. However, most airport-based businesses occupy leased property and the sale of such businesses typically reflects the transfer of leasehold interest and going-concern. Accordingly, the allocation of real estate and business value is difficult to separate. Therefore, this approach is not typically utilized.

Income Approach: This approach is based upon an estimate of the property's net income potential. The *income approach* measures the present worth of anticipated future benefits derived from property ownership. There are two methodologies which are typically utilized to determine value using the *income approach*: direct capitalization technique and discounted cash flow. The direct capitalization approach is most appropriate for airport-based properties. Under this approach, net income (which is estimated using comparable rental rates) is capitalized at an overall rate of return to arrive at an indication of value.

Acquisition

We provide a full range of services relating to the purchase of FBOs or specialized aviation businesses including:

- Assessing (and/or evaluating) potential acquisition opportunities.
- Valuing the business (or the assets of the business) to establish a purchase price.
- Providing advice regarding the structure of the transaction.
- Performing due diligence.

Once the acquisition is complete, to ensure a successful transition and in support of the on-going operation of the business, we can provide a full range of development, operation, management, and marketing related services.

Divestiture

We provide a full range of services relating to the sale of FBOs or specialized aviation businesses including:

- Valuing the business (or the assets of the business) to establish a sales price.
- Developing an offering document.
- Qualifying prospective purchasers.
- Providing advice regarding the structure of the transaction.
- Compiling a due diligence notebook.
- Coordinating and managing the due diligence process.

Lease Rate Analysis (Rates and Charges)

A lease rate analysis is used to establish market rental rates (and/or fees) for airport land and/or improvements. Utilizing this streamlined approach, rental rates for similar aeronautical land and/or improvements at similar (or comparable) airports are analyzed to derive an appropriate rental rate.

To achieve this objective, we:

- Develop a profile of the subject airport (and the leasehold-land and/or improvements).
- Identify comparable airports using the subject profile.
- Obtain rental rates (and fees) and related information from the comparable airports identified (or from our database).
- Analyze the data obtained from comparable airports.

The proper assimilation (and analysis) of comparable data (rental rates and/or fees) is the key to this process. Accordingly, we have developed and maintain an extensive database of airports (and leaseholds) specifically for this purpose.

Based upon our analysis, we recommend appropriate rental rates (and/or fees) for the land and/or improvements under study.

Strategic/Business Planning

In addition to establishing and communicating the mission, vision, and values (management philosophy and approach) of an organization, this process is utilized to develop specific goals, objectives, tactics, and action plans which (collectively) serve as a gameplan for continued and future operations.

Within this context, strategic/business plans typically encompass each of the following areas (and/or elements):

- Business Overview (historical perspective, current situation, ownership, organizational structure, key personnel, products and services, facilities and vehicles and equipment).
- Management Plan (mission, vision, and values, staffing, employee training and development, compensation and management information systems).
- Marketing Plan (market overview, situational analysis, target market identification, market positioning, market share analysis, competitive analysis and marketing/sales strategy).
- Financial Plan (budgeting, breakeven analysis, cost/expense controls, assets, and pricing strategy).

Lease and Rates and Charges Policy

In addition to setting forth the parameters for leasing land and/or improvements at an airport, this document outlines the process for establishing and adjusting rates and fees. In many respects, this document sets the stage for the other primary guiding documents (Minimum Standards, Rules and Regulations and Development Guidelines).

Operations

Operational and Financial Assessment

This process is used to ascertain how well an organization is performing in comparison to the industry (and/or comparable operations) and to identify those areas where the level, quality, and delivery of service (and/or the efficiency, cost-effectiveness, and/or safety of the operation) can be improved or enhanced. Benchmarks can also be provided (if desired). The assessment typically encompasses one or more of the following areas:

- Organizational structure
- Key personnel
- Staffing (management, administrative and operational)
- Strategic/business planning
- Products and services
- Pricing
- Training
- Management information/reporting systems
- Facilities

- Vehicles and equipment
- Financial performance
- Cost controls
- Marketing
- Business development
- Customer, community, and vendor/supplier relations.

A report identifying the most significant findings (in order of importance) and our recommendations will be provided upon completion of the process.

RFP Response Development (Proposal and Presentation)

We provide a full range of proposal development services and support including:

- Assessment of the opportunity
- Development and implementation of a response strategy
- Development and production of the proposal document and all inserts, attachments and exhibits including :
    - Technical (and creative) writing and editing;
    - Design and layout
    - Production
    - Packaging (design and production of all packaging elements); and
    - Collating and assembly
    - Final quality assurance check/inspection.
- Full coordination/management (planning, scheduling and monitoring) of all aspects of the process.

The overall objective of this process is to develop a clear, concise, easily readable, fully understandable, and completely responsive proposal in order to secure a lease, use, or operating agreement with an airport owner/operator in a desired market.

Minimum Standards

By definition, minimum standards are the "qualifications that may be established by an airport owner/operator as the minimum requirements to be met as a condition for the right to conduct an aeronautical activity on the airport." The purpose of minimum standards is "to provide a fair and reasonable opportunity, without unlawful discrimination, to all applicants to qualify, or otherwise compete, to occupy available airport land and/or improvements and engage in authorized aeronautical activities at an airport." In essence, by establishing minimum entry-level requirements (or thresholds), minimum standards level the playing field.

Rules and Regulations

Rules and regulations are established "to facilitate the safe, orderly, and efficient use of the airport." The purpose of rules and regulations are "to ensure that airport tenants and customers operate in a safe and orderly fashion and to restrict (or prevent) any activity which would interfere with the safe and orderly use of the airport."

RFP Development and Response (Proposal) Evaluation

We provide a full range of RFP development and response evaluation services including:

- Assessing the opportunity.
- Developing and implementing a solicitation strategy.
- Developing the RFP document and all attachments including technical (and creative) writing and editing, layout (design), production, and packaging.
- Advertising the opportunity and distributing the document.
- Soliciting responses from experienced and qualified prospects.
- Pre-proposal meeting preparation and debriefing.
- Developing and distributing addendums in response to questions, inquiries, and/or requests for information.
- Participating in the interview process.
- Evaluating and ranking the responses (proposals) received and/or presentations made.
- Facilitating the development of a lease, use, or operating agreement with the successful proponent.
- Coordinating/managing (planning, scheduling, and monitoring) all aspects of the RFP process.

The overall objective of this process is to develop a clear, concise, and easily readable and understood RFP document that will attract (and ultimately secure a lease, use, or operating agreement with) an experienced and qualified operator.

Litigation Support and Expert Testimony

We provide a full range of litigation support services and expert testimony in all areas involving the development, operation, management, marketing, valuation, sale, and acquisition of FBOs, specialized aviation businesses, and general aviation airports.

Lease, Use and Operating Agreements

We provide a full range of services to facilitate the development/ negotiation of lease, use, and/or operating agreements. Our expertise encompasses the following areas:

- Use of the leased premises—aviation (mandatory, optional, and additional), non-aviation, restrictions, and subletting.
- Lease term—capital investment, base term, options, renewals, and reversion of improvements.
- Rates and charges—establishing and setting, or adjusting and escalating.
- Responsibilities and obligations of both parties.
- Leasehold development—guidelines/standards and process.
- All other terms and conditions typically found in aviation/ airport lease, use, and/or operating agreements.

As a result, we are uniquely qualified to develop new leases or modify (or amend) existing leases.

AOPA

With a membership base of more than 410,000, or two-thirds of all pilots in the United States, AOPA is the largest, most influential aviation association in the world. AOPA has achieved its prominent position through effective advocacy, enlightened leadership, technical competence, and hard work. Providing member services that range from representation at the federal, state, and local levels to legal services, advice, and other assistance, AOPA has built a service organization that far exceeds any other in the aviation community.

INTERNATIONAL ASSOCIATION OF AIRPORT EXECUTIVES

The International Association of Airport Executives Canada (IAAE Canada) exists to promote the profession of airport management and the professional development of airport managers in Canada.

Since its inception in 1994, IAAE Canada has helped many members accomplish their goals by providing various educational and career enhancing opportunities.

For these reasons, IAAE Canada is proud to offer its premier credential the Accredited Airport Executive (A.A.E.), "The Mark of an Airport Professional."

The value of the A.A.E. mark is the public, the airport industry and the candidate knowing that only those who have proved a high level of knowledge, managerial skill and professionalism are awarded the Accredited Airport Executive designation.

In the early 1990s, AAAE determined that, in order to effectively address the challenges of managing airports in a global economy, there was a need for advanced airport management education and professional development worldwide—a need for a membership organization for individuals in airport management outside the United States.

Vision

To be the premier organization for professional development of airport management in Canada.

Mission Statement

The International Association of Airport Executives Canada is a member organization of airport professionals whose mandate focus is on communications and education opportunities within the industry through professional organizations and through the accreditation programme.

To respond to that need, AAAE's commitment to professional excellence and the AAAE accreditation programme went international through the creation in 1992 of the International Association of Airport Executives (IAAE), an affiliated AAAE organization.

AAAE membership is truly representative of airport management throughout the country. It places equal emphasis on large and small airport concerns, reflecting a membership comprised of executives from large and medium size airports, as well as hundreds of managers from smaller airports used exclusively by general aviation or commuter airlines.

The American Association of Airport Executives is the largest professional organization for airport executives in the world, representing thousands of airport management personnel at public use airports worldwide. AAAE's primary goal is to assist airport executives in fulfilling their responsibilities to the airports and communities they serve.

IAAE Canada Code of Ethics

Code of Ethics

As a professional Airport Executive in the performance of my duties and responsibilities, I pledge to:

1. Be dedicated to the safe, efficient and economic operation of airports and the national air transportation system and believe that professional management is the method to achieve this objective.
2. Be dedicated to the highest ideals of honour and integrity in all-public and personal relationships in order to earn and retain the respect and confidence of public officials, employees and the general public.
3. Refrain from activities that would compromise the performance of a professional executive.
4. Resist any encroachment on professional responsibility and duties, believing that the executive should be free to carry out official policies without interference and handle each problem without discrimination on the basis of principle and justice.
5. Neither seek nor accept any favour of gift from persons or firms under which circumstances could be reasonably inferred that the favour would influence or compromise the executive or governing body.
6. Submit policy proposals to the employing governing body and officials; provide them with facts and advice on matters of policy as a basis for making decisions and establishing airport goals, implement and execute policies adopted by the governing officials.
7. Recognize that elected and/or appointed officials are responsible for the establishment of airport policies and that policy execution is the responsibility of the airport executive.
8. Keep the community informed on airport affairs, encourage communication between the users, citizens and all public officials, emphasize friendly and courteous service to the public and constantly seek to improve the quality and public image of the airport executives.

9. Handle all personnel matters on the basis of merit in order that fairness and impartiality govern the executive's decision pertaining to appointments, salary adjustments, promotions and discipline.
10. Strive to continually improve professional competence and ability through education and research; and to encourage the professional development of associates and subordinates.

Canadian Airport Executive Accreditation Programme

The International Association of Airport Executives Canada (IAAE Canada) exists to promote the profession of airport management and the professional development of airport managers in Canada.

Since its inception in 1994, IAAE Canada has helped many members accomplish their goals by providing various educational and career enhancing opportunities. For these reasons, IAAE Canada is proud to offer its premier credential the Accredited Airport Executive (A.A.E.) "The Mark of an Airport Professional."

The value of the A.A.E. mark is the public, the airport industry and the candidate knowing that only those who have proved a high level of knowledge, managerial skill and professionalism are awarded the Accredited Airport Executive designation.

The IAAE Canada Accreditation programme is set to be a designed to be of a 4th Year University academic level and challenges the candidate in proving their knowledge and skills in three distinct ways.

The IAAE Canada Board of Examiners expects all candidates to produce work of very high quality and professional standard. The Board of Examiners' members when evaluating candidates' submissions will only approve those that meet the high standards of quality and professionalism as set by the IAAE Canada Board of Directors and Board of Examiners.

Prerequisites

In order to become accredited, an individual must first be an Affiliate Member of IAAE Canada. Affiliate Membership is only open to individuals whom have active responsibility for the management, operation or administration of an airport/aerodrome in Canada.

Prospective accreditation candidates must be:

- 21 years of age;
- A current affiliate member of IAAE Canada;
- Have one full year of airport management experience; and
- Possess a university degree, college diploma, or have eight years of Canadian airport experience. Candidates may substitute airport experience in lieu of university/college on a two-for-one basis.

The Process

To become an Accredited Airport Executive the candidate must:

- Successfully pass the multiple-choice Accreditation exam based on the IAAE Canada Body of Knowledge, the Canadian Aviation Regulations and other airport documents. A 1-week Accreditation Academy is offered to review the relevant material and to write the exam. The exam without the academy can be challenged, provided an approved proctor is available;
- Complete an approved management research paper;
- Successfully complete an interview and oral examination with the Accreditation Board of Examiners; and
- Accrue at least three full years of experience in airport management employment.

After the candidate has successfully completed all of the programme requirements he or she becomes an "Accredited Airport Executive" and the member is allowed the use of the A.A.E. designation after their name.

Procedure

Candidates may complete either the management research paper or the written examination first. The final phase is the interview and oral examination with the Board of Examiners. Candidates are expected to complete the programme within three years.

IAAE Canada Certified Member Programme

Affiliate Candidates and Associate Members are eligible for the Certified Member Programme.

Prerequisites

Prospective Certified Members (C.M.) must be:

- 21 years of age,

- A current member of IAAE Canada,
- Have one full year of airport management experience, and
- Possess a university degree, college diploma, or have eight years of Canadian airport experience. Candidates may substitute airport experience in lieu of university/college on a two-for-one basis.

The Process

Successfully pass the multiple-choices Accreditation exam based on the IAAE Canada Body of Knowledge, the Canadian Aviation Regulations and other airport documents. A 1-week Accreditation Academy is offered to review the relevant material and to write the exam. The exam without the academy can be challenged, provided an approved proctor is available.

After the candidate has successfully completed the programme requirements he or she becomes a "Certified Member" and the member is allowed the use of the C.M. designation after their name.

Programme Enrollment

Members interested in pursuing the Airport Executive Accreditation Programme and/or Certified Member Programme should complete the attached form and submit it along with a current resume to the IAAE Canada office. You will be sent the programme materials upon acceptance of your application and the payment of the programme fee.

If you have questions, please contact: IAAE Canada by e-mail at headquarters@iaaecanada.org.

Written Exam

The written exam is a multiple-choice exam that is developed by the IAAE Canada Accreditation Committee and the IAAE Canada Board of Examiners with input from aviation and education experts. The exam is based on the IAAE Canada Body of Knowledge (the text Airport Development, Management and Operations in Canada, by E.R. Syme and A.T. Wells), the IAAE Canada Accreditation Programme Study Guide and the Canadian Aviation Regulations.

IAAE Canada offers a 1-week Accreditation Academy that reviews the relevant material and proctors the exam. The academy is offered by IAAE Canada when demand exists.

A candidate may challenge the exam without completing the academy for a fee as set by the IAAE Canada Board of Directors. Candidates challenging the exam are required to have a proctor supervise the examination, and the IAAE Canada Accreditation Committee must pre-approve the proctor.

Management Paper

The management paper is a reflection of the professionalism within IAAE Canada and contributes to the IAAE Canada body of knowledge. The completed papers are included in the IAAE Canada library, which is accessible to all IAAE Canada, IAAE and AAAE members.

In general, papers are to be: 15-25 pages in length (exclusive of figures, indexes, appendices, etc.); relevant to airport management in Canada; and "transportable" to other airports in Canada.

All three Examiners must agree that the paper meets the prescribed measures of professionalism and format as listed in the IAAE Canada Management Paper Guidebook.

Final Interview

Before completion of the final interview, the candidate must meet the following requirements:

- Have successfully completed both the written examination and management research paper phases of the accreditation programme;
- Have accrued at least three full years of Canadian airport management experience; and
- Sign a code-of-ethics form to be kept on record at the IAAE Canada offices.

The format and process of the interview, along with the interview questions can be found in the IAAE Canada Interview Panel Members Guide. Guidelines for the candidates are in the IAAE Canada Candidates Guide to the Final Interview.

The Canadian Board of Examiners

- *Cuyler Green,* A.A.E., Chair Board of Examiners
- *Bill Newman,* A.A.E.
- *Bryan Gordon,* A.A.E.
- *Dave Kerrigan,* A.A.E.

- *David Hunter*, A.A.E.
- *Eric Tolton*, A.A.E.
- *Eugene McDonald*, A.A.E.
- *Lloyd McCoomb*, A.A.E.
- *Roger Sellick*, A.A.E.
- *Shawn Sutherland*, A.A.E.
- *Steve Bruchi*, A.A.E.
- *Stieg Hoeg*, A.A.E.

Current Fee Structure

- Joining the Programme...... ....$ 300 + GST
- Writing the Exam...................$ 200 + GST
- Academy Enrollment..........$ 1,000 + GST (Includes the exam cost)

FEDERAL AVIATION INTERACTIVE REPORTING SYSTEM (FAIRS)

FAIRS is a highly secure, web-based aviation management information system that GSA uses to collect and analyze the costs and usage (in hours) of the aircraft that the federal agencies own or hire (i.e., rent, charter, lease, full-service and other contracts). Annually, GSA submits a report from FAIRS to the Office of Management and Budget (OMB).

FAIRS is Online

GSA deployed FAIRS on 1 April, 2000, and the system is available for use by all federal government agencies that own or hire aircraft. FAIRS is an Internet-based system that federal government users access through a highly secure website. Each user has his/her own digital certificate to ensure security, and every FAIRS transaction is encrypted.

GSA has developed and operates FAIRS in compliance with Office of Management and Budget (OMB) Circular A-126, "Improving the Management and Use of Government Aircraft." Circular A-126 requires GSA to maintain a management information system to collect information about the numbers, cost, and usage of aircraft that federal agencies own or hire.

Quarterly, each federal government agency is responsible for reporting data to FAIRS on the costs and usage of its owned (i.e., federal) aircraft. Also each quarter, agencies must report the costs and usage of their Commercial Aviation Services (CAS). Within two weeks of making any change to their federal aircraft inventory, an agency must report the change to FAIRS.

In spring 2005, GSA published the fifth annual report from FAIRS, "The Status of Federal Government Aircraft in Fiscal Year 2005." This report contains extensive data on the aircraft owned and hired by the U.S. federal government. More information on the FAIRS Report is available from Jay Spurr ('FAIRS' Programme Manager).

FAIRS Operation: GSA operates FAIRS under business rules that spell out the basic assumptions and requirements for reporting and for maintaining the system's security. Members of the Interagency Committee for Aviation Policy (ICAP) (Management Data and Systems Sub-committee) worked with GSA to develop the business rules for FAIRS and continue to serve as the change control board for the system.

Who can use FAIRS? Only users and reviewers chosen by their agencies and trained by GSA can access the FAIRS' Website. GSA assigns digital certificates with special reference numbers and passwords to each authorized user and reviewer.

The Query Tool: In addition to the basic FAIRS application and data base, the system includes a query tool (Business Objects), which allows the federal agencies, GSA, and OMB to run a series of prepared reports or to query the FAIRS data base on an ad hoc basis. These capabilities give the federal community, significant flexibility in deriving useful information for aviation planning, programming and budgeting.

Other Management Tools: Along with FAIRS, GSA provides a number of helpful management tools, which include Jane's Aircraft Directory, the World Aviation Directory, the Aircraft Blue Book, the ICAP Cost Accounting Guide, the Conklin and deDecker Life-Cycle Cost Analyzer, and a set of aviation clip art.

The current 'FAIRS' system will be replaced in February, 2007. 'Entrust' and 'Business Objects' will no longer be used, nor will Oracle comprise the application database. The new 'FAIRS' will be hosted and housed at the Idaho National laboratory in Idaho Falls, Idaho. The old system will be turned off 12 January, 2007. The new system will be turned on 5 February, 2007.

AIRCRAFT REGULATIONS AND GUIDANCE

GSA publishes regulatory policy for aircraft management in 41 Code of Federal Regulations (CFR) 102-33, "Management of Government Aircraft," and 41 CFR 300-3; 301-10; and 301-70, "Travel on Government Aircraft." The latest version of 41 CFR 102-33 was published in the Federal Register on 6 November, 2002. The latest policy for "Travel on Government Aircraft" was published 21 June, 2004.

Office of Management and Budget Circular A-126, "Improving the Management and Use of Government Aircraft," revised May, 1992, provides the basic guidance for management of federal aviation programmes and for travel on government aircraft. 41 CFR 102-33 and 41 CFR 300-3; 301-10; and 301-70 codify the guidance in OMB Circular A-126.

GSA also publishes a number of other guides and manuals to help federal agencies manage the acquisition, use, and disposal of their aircraft (see the "Other Federal Aviation Guidance" link). These publications include, the "U.S. Government Aircraft Cost Accounting Guide," revised November, 2002, which contains essential guidance for accounting for government aircraft costs, for justifying the use of government aircraft, and for recovering costs.

The "Safety Standards Guidelines for Federal Flight Programmes" are high-level guidelines for the federal agencies to use to create their own agency-specific standards. Fifteen federal agencies voluntarily adopted these guidelines by signing the Interagency Committee for Aviation Policy (ICAP) "Safety Standards Agreement" in 2000-01.

Federal agencies may fly their aircraft as "public aircraft" under certain conditions, described in Public Law 106-181 and the Federal Aviation Administration's (FAA) Advisory Circular on Public Aircraft.

CONVENTION ON INTERNATIONAL CIVIL AVIATION

The **Convention on International Civil Aviation**, also known as the **Chicago Convention**, established the International Civil Aviation Organization (ICAO), a specialized agency of the United Nations charged with coordinating and regulating international air travel. The convention establishes rules of airspace, airplane registration and safety, and details the rights of the signatories in relation to air travel.

The document was signed on 7 November, 1944 in Chicago, Illinois, by 52 signatory states. It received the requisite 26th ratification on 5 March, 1947 and went into effect on 4 April, 1947. ICAO came into being on 4 April, 1947. In October of the same year, ICAO became a specialized agency of the United Nations linked to Economic and Social Council (ECOSOC). The convention has since been revised eight times (in 1959, 1963, 1969, 1975, 1980, 1997, 2000 and 2006). Links to all versions of the document can be found in the external links section.

The original signed document resides in the National Archives of the United States.

Convention on International Civil Aviation (also known as *Chicago Convention*), was signed on 7 November, 1944 by 52 States. Pending ratification of the Convention by 26 States, the Provisional International Civil Aviation Organization (PICAO) was established. It functioned from 6 June, 1945 until 4 April, 1947. By 5 March, 1947 the 26th ratification was received. ICAO came into being on 4 April, 1947. In October of the same year, ICAO became a specialized agency of the United Nations linked to Economic and Social Council (ECOSOC).

The *Convention on International Civil Aviation* set forth the purpose of ICAO:

"WHEREAS the future development of international civil aviation can greatly help to create and preserve friendship and understanding among the nations and peoples of the world, yet its abuse can become a threat to the general security; and

WHEREAS it is desirable to avoid friction and to promote that co-operation between nations and peoples upon which the peace of the world depends;

THEREFORE, the undersigned governments having agreed on certain principles and arrangements in order that international civil aviation may be developed in a safe and orderly manner and that international air transport services may be established on the basis of equality of opportunity and operated soundly and economically;

Have accordingly concluded this Convention to that end."

INTER-AGENCY COMMITTEE FOR AVIATION POLICY (ICAP)

Federal agencies that own or hire aircraft are members of the ICAP, which GSA established in 1989 at the direction of the Office of

Management and Budget (OMB). The ICAP advises GSA on making policy for federal aviation management.

Through the ICAP, GSA and federal agencies work together to identify and coordinate the policy views of the federal aviation community and to foster the safest, most efficient and effective federal aviation operations. GSA chairs and facilitates the committee, provides programmes and tools to support aviation activities, and operates a management information system to collect and report data related to federal aviation management. The ICAP accomplishes its objectives through four sub-committees: Acquisition, Use, and Disposal Sub-committee, Communications Sub-committee, Management Data and Systems Sub-committee, and Safety Standards and Training Sub-committee.

Members of the ICAP

- Department of Agriculture
- Department of Commerce
- Department of Defence
- Department of Energy
- Department of Health and Human Services
- Department of Homeland Security
- Department of Justice
- Department of State
- Department of the Interior
- Department of the Treasury
- Department of Transportation
- Department of Veterans Affairs
- Environmental Protection Agency
- Federal Aviation Administration
- Federal Emergency Management Agency
- General Services Administration
- National Aeronautics and Space Administration
- National Science Foundation
- National Transportation Safety Board
- Office of Management and Budget
- Tennessee Valley Authority.

GSA makes policies for managing the acquisition, use, and disposal of aircraft that the federal civilian agencies own or hire. In addition, GSA does the following:

- Sponsors the Interagency Committee for Aviation Policy (ICAP).

- Collects, analyzes, and reports information on government aircraft, using the Federal Aviation Interactive Reporting System (FAIRS).
- Promotes best practices in federal aviation management.

In cooperation with ICAP-member agencies, GSA's overarching goal is to foster the safest, most effective and efficient aviation in U.S. government agencies.

ACTIVE LEVEL OF SERVICES REVIEWS

In 2003 NAV CANADA conducted a review of services the company provides across the country. Airport control and advisory services, weather services, electronic navigational aids and approaches, emergency navigation assistance and communications services were included in the review.

The purpose of the level of service review is to ensure NAV CANADA's services match our customer's requirements. Safety was our first priority in examining any potential changes. We also sought to not impact customer efficiency adversely, and to take into consideration the future system requirements and the ability of the air navigation system to adapt to changing customer needs. The review identified the need for service enhancements and the opening of new facilities and the installation of new equipment as well as service reductions and facility closures.

The preliminary analysis of the level of service review were originally released in November, 2003 for input. Following consideration of the input received, proposals for changes in service were then released in January, 2004.

An aeronautical study is currently underway to examine the proposals for level of service change. The study process involves consulatation, which will take place between February and July, 2004.

GSA has developed an extremely successful Airline City Pair Programme. This service originally covered only 11 markets, but over the last 20 years, it has expanded to over 4,000 city pairs. The airfares offered under this programme are discounted considerably off comparable commercial fares—saving the federal government billions of dollars annually.

In addition to the tremendous price savings, the Airline City Pair Programme has many features which allow government travelers all the flexibility possible in planning official travell. The benefits of this service include:

- Fares priced on one-way routes permitting agencies to plan multiple destinations;
- No advance purchase required;
- No minimum or maximum length stay required;
- Tickets fully refundable;
- Last seat availability;
- No blackout periods;
- Stable prices enabling travel budgeting; and
- Dual fares availability.

AIRCRAFT ENGINEERS INTERNATIONAL (AEI) AFFILIATION

AMFA National has called for a referendum vote regarding an Affiliation with Aircraft Engineers International (AEI) in accordance with Article XI, Section 1 of the AMFA Constitution. In an effort to aid the AMFA membership in making an informed decision regarding an affiliation with the Aircraft Engineers International (AEI), AMFA National has provided links to reference information at the bottom of this posting.

Votes for this referendum will be cast using the TrueBallot Inc., WebVote® and TeleVote® systems. All active AMFA members in good standing are eligible to vote in this referendum.

Note: Any member whose membership dues, fees, and/or assessment obligations have not been paid in full is a member in bad standing and, therefore, their vote will be cancelled in this referendum. Members are strongly urged to resolve any arrearage prior to this referendum being conducted.

The referendum notice and voting instructions were mailed to eligible voters at the address on file with AMFA National as of 27 February, 2007. Any eligible member who has not received voting instructions in the mail by 15 March, 2007, or anyone having questions about the voting process, should contact the AMFA Election Administrator at 720-744-6629. The voting period for this referendum will commence on 27 February, 2007 at 3:00 p.m. ET and will end on 29 March, 2007, at 10:00 a.m. ET.

BCAS INDIA: A CASE HISTORY

History

The Bureau of Civil Aviation Security (BCAS) was initially set-up as a Cell in the Directorate General of Civil Aviation (DGCA) in

January, 1978 on the recommendation of the Pande Committee constituted in the wake of the hijacking of the Indian Airlines flight on 10th September, 1976. The role of the Cell was to coordinate, monitor, inspect and train personnel in Civil Aviation Security matters.

The BCAS was reorganized into an independent department on 1st April, 1987 under the Ministry of Civil Aviation as a sequel to the Kanishka Tragedy in June, 1985. The main responsibility of BCAS are lay down standards and measures in respect of security of civil flights at International and domestic airports in India.

Organisation

- BCAS is the regulatory authority for civil aviation security in India. It is headed by an officer of the rank of Director General of Police and is designated as Commissioner of Security (Civil Aviation).
- Commissioner of Security (CA) is the appropriate authority for implementation of Annexure 17 to Chicago Convention of International Civil Aviation Organization (ICAO).
- Commissioner of security (CA) is responsible for the development, implementation and maintenance of the National Civil Aviation Security Programme.
- BCAS Hqrs is located at **"A" Wing, I-III floor, Janpath Bhavan, Janpath, New Delhi-110001.** It has got four Regional Offices located at International airports i.e. Delhi, Mumbai, Kolkata and Chennai. The Regional Office is headed by an officer of the rank of Deputy Commissioner of Security (CA).

Functions

- Laying down Aviation Security Standards in accordance with Annexure 17 to Chicago Convention of ICAO for airport operators, airlines operators, and their security agencies responsible for implementing AVSEC measures.
- Monitoring the implementation of security rules and regulations and carrying out survey of security needs.
- Ensure that the persons implementing security controls are appropriately trained and possess all competencies required to perform their duties.
- Planning and coordination of Aviation security matters.

- Conducting:
  - Surprise/Dummy checks to test professional efficiency and alertness of security staff.
  - Mock exercise to test efficacy of Contingency Plans and operational preparedness of the various agencies.

Central Industrial Security Force (CISF)

Security functions at all the airports in the country were performed by the police personnel requisitioned from State Governments till January, 2000. In the backdrop of hijacking of Indian Airlines aircraft (IC-814) in December, 1999, airport security matters were reviewed by the Ministry of Civil Aviation. It was decided that in order to bring in uniformity of practices and procedures and ensure effective control and supervision of the Ministry of Civil Aviation, airports security should be entrusted to a single dedicated force instead of different State Police forces with divergent work culture and practices. The Committee of Secretaries (COS) in its meeting on 7 January, 2000 recommended that in the long-term there was a need for a more professionalised force for civil aviation. The Commissioner of Security (CA) further recommended that Central Industrial Security Force (CISF) should be inducted at all airports in India. The dedicated CISF contingent earmarked for aviation security functions at airports in India has been notified as Aviation Security Group (ASG).

After deployment of CISF at airports, there has been a marked improvement in all areas of airport security including Perimeter Security, Access Control, Terminal Building Security, Apron Security, Surveillance, Passenger Handling etc. ASG staff have proved their worth in handling all types of security situations at airports and have been meticulous in implementing security procedures. They have been quick to plug any loopholes in security and the security system is constantly upgraded to neutralize emerging threats. In addition they have been able to bring about a higher level of security awareness among all agencies operating at the airports, including the passengers.

Aviation Security Group of CISF is passenger friendly without compromising on security standards. ASG is committed to providing a safe and secure environment at all the airports under its charge. The dual objective of the ASG is to ensure that air travel is absolutely safe and the passengers carry a pleasant experience of airport security.

AVSEC RULES AND REGULATIONS

Sl. No.	**ACTS**
1.	Aircraft Act, 1934
2.	Aircraft Rules, 1937
3.	Tokyo Convention 1975 Act
4.	Anti-Hijacking Act, 1982
5.	Anti- Hijacking Act, 1994 (Amendments 1994)
6.	The Suppression of Unlawfull Acts Against Safety of Civil Aviation Act, 1982
7.	The Suppression of Unlawfull Acts Against Safety of Civil Aviation Act , 1994.

AIR TRAVEL IN INDIA: AN OVERVIEW

Due to introduction of enhanced security measures based on threat assessment across the country, BCAS continues to improve and redefine security procedures to make it secure and passenger friendly.

Useful tips and informative material given below will ensure a smooth passage thorough security at any of the civil airports in India:

- *Travel Tips*: A **must read** for anyone travelling by air.
- *Travel Safe and Smart*: Necessary preparations can be made by you before you arrive at the airport.
- *Permitted and Prohibited Items*: You could be surprised at the items, which are permitted and prohibited on-board an aircraft.
- *Transporting Special Items*: Some helpful information that explains the security screening procedures as they may apply to unique items.
- *Travelling with Children*: Everything you need to consider when travelling with children.
- *Special Considerations*: Covers numerous topics such as film, pets, religious/cultural needs, sporting equipment, etc.—Covers numerous topics such as film, pets, religious/cultural needs, sporting equipment, etc.

- *Persons with Disabilities and Medical Conditions*: This area is extensive. BCAS has divided the area into groups and has trained the screener workforce appropriately.
- *Security Awareness*: Learn what to look for, what to do.
- *Frequently Asked Questions*: Answers to your most Frequently Asked Questions.—Answers to your most Frequently Asked Questions.

Travel Safe and Smart

Preparations

Necessary preparations can be made by you before arrival at the airport which will help you to move more quickly and efficiently through the security processes. Here you will find suggestions on what to wear to the airport and how to pack for your trip. We've also included a pre-flight checklist to help you to travel safe and smart.

Dress

Security does not require any particular style or type of clothing. However, certain clothing and accessories can set off an alarm on the metal detector and may affect the pax flow through the screening points. Here you will find tips to help you.

Pack Smart

There are restrictions on what you can pack in your hand baggage and registered baggage. All of your baggage will be screened and possibly hand-searched as part of the security measures. This inspection may include emptying most or all of the articles in your bag. Here you will find tips to help you pack.

Final Checklist

You're dressed, packed and ready to go. Here is a pre-flight checklist to help you to travel safe and smart. Read the instructions printed on the air ticket and contact your airline or travel agent for additional information.

Access Requirements

You can enter into the passenger terminal if you have either a confirmed air ticket for the journey or a valid airport entry permit.

Visitors can buy ticket to enter into the visitor's area in the terminal. Waitlisted passengers are advised to contact the airlines office on the landside of the terminal for confirmation of their tickets.

Security Process and Procedures

Familiarise yourself with the Security Process and Procedures. It will help you to play an active role in ensuring your own safety and avoid inconvenience to yourself.

Transporting Special Items

Bureau of Civil Aviation security has developed standardized security screening procedures for all airports in the country. Therefore, you can expect that you will encounter essentially the same procedures at each airport you visit.

We have identified some helpful information that explains the security screening procedures as they may apply to unique items:

- Alcoholic Beverages
- Camping Equipment
- Deceased Remains
- Fire Arms and Ammunition
- Hunting and Fishing
- Musical Instruments
- Parachutes
- Photographic Equipment and Film
- Scuba Equipment
- Sporting Equipment.

Note: Please check with your airline before arriving at the airport. Security staff will screen any item regardless of type or size.

As long as it can be cleared, it will be allowed through the checkpoint. It is the air carrier's responsibility to restrict items due to size limitations and these standards are different from one air carrier to another.

Your knowledge of the procedures, and observance of the tips and recommendations, will help you move through the security checkpoint quickly and efficiently.

This information will be updated from time-to-time. We recommend that you periodically check this website to obtain the latest information.

Travelling with Children

Every person, regardless of age, must undergo screening to proceed beyond the security checkpoint. Even babies must be individually screened. **You will not be asked to do anything that will separate you from your child or children.**

Screeners are specially trained and understand your concerns regarding children. Your children will be approached gently and treated with respect. If your child becomes uncomfortable or upset, you will be consulted about the best approach to resolving your child's concern.

The Screening Process

X-Ray

- All carry-on baggage, including children's bags and items, must go through the X-ray machine. Examples include: diaper bags, blankets and toys.
- All child-related equipment that will fit through the X-ray machine must go through the X-ray machine.
- Strollers, umbrella strollers, baby carriers, car and booster seats, backpacks and baby slings.
- When you arrive at the checkpoint, you should collapse or fold child-related equipment and secure items that are in the pockets, baskets, or attached to the equipment. You will place these items on the X-ray belt for inspection. Plastic bins are provided to deposit such items.
- When child-related equipment does not fit through the X-ray machine, the equipment must be visually and physically inspected.
- Ask screeners for assistance to help reunite you with your bags and child-related equipment, if needed.

Alert!

Babies should **NEVER** be left in an infant carrier while it goes through the X-ray machine. For information regarding what is permitted or prohibited from being in carry-on luggage, please refer to our prohibited items section.

The Walk through Metal Detector/Door Frame Metal Detector

The screener will need to resolve the alarm for both the adult and child if something sets off the alarm as you are carrying a child through the metal detector.

- Babies and children must be removed from their strollers or infant carriers so they can be individually screened.
- You may not pass the child to another person behind you or in front of you during this process.
- Do not pass your child to the screener to hold.
- The screener may ask for your assistance with secondary screening of your child.
- If your child can walk unassisted, it would be best to have the child walk through the metal detector independently.

Children with a Disability

If your child has a disability, screeners may ask you what abilities your child has in order to determine the best method for screening (e.g. carry the child through the walk–through metal detector, hand-wand procedure).

If your child has a disability, screeners will never attempt to remove the child from his or her equipment. This will only be done at the discretion of the accompanying adult.

TIPS Before You Reach the Airport

- Please allow yourself and your family extra time to get through security—especially when travelling with younger children.
- Call your airline or travel agent for information on recommended check-in times for your departure airport.
- Talk to your children before coming to the airport and let them know that it's against the law to make threats such as, "I have a bomb in my bag." Threats made jokingly (even by a child) can result in the entire family being delayed and could result in fines.

TIPS at the Airport

- Speak to your children again about the screening process so that they will not be frightened or surprised. Remind them not to joke about threats such as bombs or explosives.

- Advise your children that their bags (backpack, dolls, etc.) will be put in the X-ray machine and will come out at the other end and be returned to them.
- Let your children know that a screener may ask to see Mom or Dad's shoes, but that these too will be returned after the inspection.
- You may want to consider asking for a private screening if you are travelling with more than one child.

Special Considerations

Bureau of Civil Aviation Security has developed standardized security screening procedures for all airports in the country. Therefore, you can expect that you will encounter essentially the same procedures at each airport you visit.

While the same screening procedures are used for virtually all passengers, we recognize that some passengers may have special needs or require additional assistance during the screening process. To maintain excellent security and customer service, the security staff have been trained to be sensitive to and respectful of the needs of all passengers.

The Security Process and Preparation Tips

We have identified some helpful information that explains the security screening procedures as they may apply to groups with special needs. If you fall into any one of the categories identified below, you may want to familiarize yourself with this information before arriving at the airport.

- Persons with disabilities or medical concerns
- Religious or cultural needs
- Pets
- Film.

Your knowledge of the procedures, and observance of the tips and recommendations, will help you move through the security checkpoint quickly and efficiently.

This information will be updated from time-to-time. We recommend that you periodically check this website to obtain the latest information.

Persons with Disabilities and Medical Conditions

In order to assist persons with disabilities or medical concerns, BCAS has developed guidelines with input from various groups. We ensure that airport screeners adopt the methodology needed to ensure secure and comfortable screening at all checkpoints.

Please refer to the links below to assist you as you make travel preparations.

- Before You Go
- Tips for the Screening Process
- Mobility Disabilities
- Hearing Disabilities
- Visual Disabilities
- Hidden Disabilities
  - —Pacemakers, Defibrillators, Other Implanted Medical Devices, and Metal Implants
  - —Diabetes
  - —Medications
- Assistive Devices and Mobility Aids
  - —Prosthetic Devices, Casts, and Body Braces
  - —Walkers, Crutches, and Canes
  - —Orthopedic Shoes, Support Appliances and Exterior Medical Devices
- Service Animals
- Children with Disabilities
- Medical Conditions and Special Situations

Security Awareness

Security awareness is everyone's responsibility. To ensure that your travel is safe, secure and efficient, take time to make security awareness an integral part of your travel experience.

General Guidelines

- Be aware of suspicious activity in and around your immediate surroundings. Inform the proper authorities if you are concerned about an unattended item or suspicious activity in and around any facility. Be extra careful if you work on your laptop. Working takes your attention away from your surroundings:
  - —Describe suspicious device(s). Visual check only. Do NOT touch object.

—Do NOT use your cell phone within 50 feet of a suspicious item.

- Remember to control all carry-on bags and keep your items and tickets with you at all times. Medicines, cash, jewellery, business papers and valuables should be packed in carry-on bags.
- Never leave anything unattended as it could be subjected to tampering and/or theft.
- Never carry anything for another person, especially for someone you do not know.

For more information on security, please refer to the resources section or contact your transportation provider directly.

Frequently Asked Questions (FAQs)

Primary Q's and A's

Q: *What are Permitted and Prohibited Items?*

A: Prohibited items are weapons, explosives, incendiaries, and items that are seemingly harmless but may be used as weapons—the so-called "dual-use" items. Please refer to the BCAS web page for more information.

Q: *How early should I arrive at the airport?*

A: Check with your airline or travel agent to determine how early to arrive at the airport. Recommended check-in times differ by airline/airport and change during peak travel times.

Q: *How should I pack before my flight?*

A: There are different restrictions on what you can pack in your carry-on and checked baggages. All of your baggage will be screened and possibly hand-searched as part of the new security measures. Please refer to BCAS **Travel safe and smart** web page to find tips to help you pack.

Q: *What can I do if my property is lost, missing or damaged during the baggage screening process or at the Security Checkpoint?*

A: It is our job to assure you—the passenger—that BCAS/Airlines/Aviation Security Group can and will deliver excellence in security and customer service.

Q: *What do I need to know if I am a Traveller with a Disability or Medical Concerns?*

A: In order to assist persons with disabilities or medical concerns, BCAS has developed guidelines with input from various groups and have trained our airport screeners in the methodology needed to ensure secure and comfortable screening at all Security Checkpoints. Please refer to the Persons with Disabilities and Medical Concerns section of the BCAS website for more information.

Checked Baggage Q's and A's:

Q: *Should I lock my luggage?*

A: Yes. In case of the requirements of physical check on doubt, you will be requested to open the baggage yourself.

Q: *How do I transport a Firearm? Ammunition?*

A: There are specific regulations governing the transport of firearms, firearm parts, and ammunition. Please refer to BCAS list of Transporting Special Items for more information.

Security Checkpoints Q's and A's:

Q: *What should I expect at the Security Checkpoint?*

A: The Passenger Security Checkpoint includes three (3) primary steps you may want to become familiar with. Please refer to Passenger Security Checkpoints for more information.

Q: *What may happen to me if I bring a prohibited item to a security checkpoint?*

A: If you bring a prohibited item to the checkpoint you may be criminally prosecuted or at the least asked to rid yourself of the item. Please refer to the list of Permitted and Prohibited Items for more information.

Q: *Can I carry a kirpan in the passenger cabin?*

A: Yes. If you are a Sikh passenger and travelling in domestic flight in India. The size of kirpan should not be more than 6 inch blade and 3 inch handle.

Q: *What is the permissible size of hand baggage allowed for air passengers?*

A: 115 cm total of all dimensions except at Srinagar airport where carry-on (hand baggage) is not allowed.

Travel Tips

A MUST read for anyone Travelling by air following tips will help you reduce your wait time at the security checkpoint:

- **Do NOT** pack or bring prohibited items to the airport. Read the **Permitted and Prohibited Items.**

- Refrain from carrying unverified gifts or presents in wrapped package. If the package alarms, screener will need to unwrap it to investigate the source of the alarm.
- **Shoes, clothing items and other accessories that contain metal** will alarm the metal detector. As a result the screener will require you to undergo further checks which may include pat down frisking.
- **Put all undeveloped films and cameras with film in your Cabin (carry-on) baggage.** Checked baggage screening equipment may damage undeveloped films.
- Carry-on baggage is limited to one carry-on bag plus one personal item. Personal items include laptops, purses, small backpacks, briefcases, or camera cases. Remember, 1+1.
- Place identification tags on all your baggage. **Don't forget to label your laptop computer.** These are one of the most forgotten items at Screening Checkpoints.
- Refrain from packing valuable items in your checked baggage (Regd. Baggage). Once you hand your baggage to your air carrier, security staff of the air carrier will handle and process the baggage in the prescribed manner Please keep in mind that most air carriers have limited liability for lost, damaged, or stolen items.
- Declare fire arms and ammunition to your airline and place it in your checked/registered baggage.
- Protect yourself and do not pack valuables in your checked baggage.

Your Cabin/Carry-on Baggage

- Carry all metal items in your carry-on bag. This includes jewelry, loose change, keys, mobile phones, pagers, and personal data assistants (PDAs).
- Take your laptop and video camera out of their case and place in the tray provided at the security checkpoint.
- Place your overcoat or jacket in the tray at the security screening Checkpoint. Suit jackets and blazers need not be removed, unless requested by the screener.
- Do not Leave your Baggage Unattended.
- Do not accept baggage from strangers. It may contain prohibited items or dangerous goods.

3

Commercial Aviation Management Programmes, Education and Training

AVIATION MANAGEMENT: A CASE STUDY OF PROGRAMME OFFERED

Programme Description

The aviation management major provides students with the skills needed for careers as managers or senior administrators for an airport, airline, government agency dealing with air commerce, or aviation-related business. The programme develops expertise in business management with emphasis on administration of various aviation enterprises.

Visit JU's Aviation Internet Resources page, a guide to over 400 resources on the internet.

Curriculum Overview

Students majoring in aviation management must complete the University core curriculum and the College of Business core requirements. The curriculum includes such areas as business economics, data acquisition and analysis, finance, marketing, and the study of aviation business law and regulations. The programme also offers applied management courses dealing with the operation of airlines, airports, and aviation businesses. Elective courses in flight training are available at the student's option.

Academic Preparation

The aviation management major has no requirements for admission beyond those required for admission to the University. However, a minimum 2.0 grade point average in courses taken for the major must be maintained for designation, continuation, and graduation as an aviation management major.

Facilities and Resources

The aviation management major is administered by the Aeronautics Programme in the Davis College of Business. The Davis building is the most modern facility on the campus. Its two levels contain classrooms, faculty offices, lounges, seminar rooms, two auditoriums, and three computer technology classrooms/labs. The aeronautical science classrooms are equipped with state-of-the-art, computer-based multimedia and simulation equipment.

Students in the programme also are exposed to unparalleled commercial aviation resources. Jacksonville University is ideally located at the heart of a major transportation hub and trade center with a diverse array of airports, air traffic control facilities, navigational facilities, and military installations. Unique to such a university programme, JU has joined forces with Delta Connection Academy, owned and operated by Delta Airlines, to provide high quality flight training to its aviation students. A leader in the regional airline industry, Delta Connection Academy has hiring agreements with the largest regional airlines including Atlantic Southeast Airlines, American Eagle, Chautauqua Airlines, Comair Airlines and Trans States Airlines.

Internship and Career Opportunities

Students have many opportunities to get a head start in their careers by participating in aviation internships and aviation career workshops. Students can intern in airline operations, airport management, airport business operations, aviation law, and FAA flight training operations. The internship programme includes sponsors such as American Airlines, COMAIR Airlines, Comair Aviation Academy, Northwest Airlines, US Airways, Jacksonville Port Authority, Denver International Airport, Paine/Boeing Field, and NASA Aviation/Space Camp. Upon successful completion of an internship, students receive academic credits toward graduation. To participate in the internship programme, students must be juniors or seniors with a cumulative GPA of at least 2.5 and a minimum 3.0 average in their major field.

Students also receive hands-on advice to plan their careers during JU's annual Aviation Career Workshop. This four-hour workshop updates students on the state of hiring in the aviation industry, provides them with advice on writing resumes and self-marketing techniques, and teaches them how to plan for long-term career goals in aviation.

Sponsored Student Activities

JU is an institutional member of the University Aviation Association, the Council on Aviation Accreditation, and the Florida Aviation Trade Association. Membership entitles students to aviation scholarships, training and career resources, and attendance at special annual events and activities.

Students also have the opportunity to develop leadership and teamwork skills by joining student chapters of Alpha Eta Rho international aviation fraternity, the American Association of Airport Executives, and the National Intercollegiate Flying Association. Through these organizations, students can meet peers who share similar interests in aviation and participate in exciting activities, such as visiting local air traffic control facilities and airports. These student organizations are very proactive in providing members with scholarships, internships and career opportunities.

Outcomes

JU aviation graduates have gone on to successful careers with commercial air carriers and other aviation-related agencies. Some of their employers include COMAIR Airlines, Dallas-Fort Worth Airport, Delta Airlines, Federal Aviation Administration, Federal Express, Florida Gulf/Mesa Airlines, Jacksonville Port Authority, Naval Air Systems Command, Northrop Grumman, Philadelphia Port Authority, United Airlines, US Airways and the U.S. Navy.

Contacts

For additional information, you can write, call, or e-mail us:

Office of Admissions 904-256-7000 or 800-225-2027 (toll-free) E-mail: admissions@ju.edu.

AVIATION MANAGEMENT AND FLIGHT OPERATIONS

Programme Description

The aviation management and flight operations major prepares candidates for careers as professional pilots for airlines or corporate flight departments. Jacksonville University has joined forces with Delta Connection Academy (DCA), owned and operated by Delta Airlines. The JU–DCA partnership combines, for the first time in college aviation education, the academic excellence of a private university with an airline-designed, airline-operated flight academy. Delta Connection Academy has hiring agreements with the largest regional airlines, including Atlantic Southeast Airlines, American Eagle, Chautauqua Airlines, Comair Airlines and Trans States Airlines.

Students benefit from a well-rounded education with a solid foundation in business management and receive airline-quality training. This unique opportunity allows students to obtain a four-year college degree in a high quality academic environment and pursue the best possible path toward a career as professional pilots.

Curriculum Overview

Students majoring in aviation management and flight operations must complete the University core curriculum and specialized College of Business core requirements. In addition, students complete courses dealing with the conduct of commercial flight operations and management of airlines, airports, and aviation-related businesses. The flight training programme, an integral part of the curriculum, provides all required FAA flight ratings to conduct flight operations as commercial pilots and flight instructors.

Academic Preparation

In addition to JU admission requirements, students planning to enroll in aviation management and flight operations must meet FAA medical certification standards. A valid FAA Class III medical certificate is required to enroll in AO 111 (Private Pilot Lab), and a valid FAA Class II medical certificate is required to enroll in any other flight courses offered in the major, with the exception of AO 315 (Glider Pilot Laboratory). In addition, a minimum 2.0 grade point average in courses taken for the major must be maintained for designation, continuation, and graduation as an aviation management and flight operations major.

Facilities and Resources

The aviation management and flight operations major is administered by the Aeronautics Programme in the Davis College of Business. The Davis building is the most modern facility on the campus. Its two levels contain classrooms, faculty offices, lounges, seminar rooms, two auditoriums, and three computer technology classrooms/labs. The aeronautical science classrooms are equipped with state-of-the-art, computer-based multimedia and simulation equipment.

Students in the programme have unparalleled exposure to commercial aviation resources and enjoy pleasant flying weather throughout the year. Jacksonville University is ideally located at the heart of a major transportation hub and trade center with a diverse array of airports, air traffic control facilities, navigational facilities, and military installations. Delta Connection Academy, headquartered in Sanford, Florida, has established a satellite base at Craig Municipal Airport, located only six miles from the main campus, to serve exclusively as JU's flight center.

Internship and Career Opportunities

Once students complete the flight training programme, they may qualify for positions as paid flight instructors and have the opportunity to log up to 1,000 hours of pilot-in-command time. After completing the phase as instructors, JU graduates are guaranteed a job interview with any of the regional airlines associated with DCA. Students have many opportunities to get a head start in their careers by participating in aviation internships and aviation career workshops. Students can intern in airline operations, airport management, airport business operations, aviation law, and FAA flight training operations. The internship programme includes sponsors such as American Airlines, COMAIR Airlines, Delta Connection Academy, Delta Airlines, Northwest Airlines, Southwest Airlines, Jacksonville Port Authority, Denver International Airport, and Paine/Boeing Field. Upon successful completion of an internship, students receive academic credits toward graduation. To participate in the internship programme, students must be juniors or seniors with a cumulative GPA of at least 2.5 and a minimum 3.0 average in their major field. In addition, students may complete the Airlines Training Orientation Programme (ATOP), an introductory airline training course, at Continental Training Center in Houston, Texas. ATOP utilizes Boeing 737 systems, cockpit procedure training, and full-motion simulation flights. Students also receive hands-on advice to plan their careers during JU's annual Aviation Career Workshop. This four-hour workshop updates students on the state of airline hiring, provides advice on writing resumes and self-marketing techniques, and teaches how to plan for long-term career goals in aviation.

Sponsored Student Activities

JU is an institutional member of the University Aviation Association, the Council on Aviation Accreditation, and the Florida Aviation Trade Association. Membership entitles students to aviation scholarships, training and career resources, and attendance at special annual events and activities. Students also have the opportunity to develop leadership and teamwork skills by joining student chapters of Alpha Eta Rho international aviation fraternity, the American Association of Airport Executives, and the National Intercollegiate Flying Association. Aeronautical students also have opportunities to meet peers who share similar interests in aviation and participate in exciting activities, such as flying the F/A-18 dome simulator at a nearby Naval Air Station or visiting local air traffic control facilities and airports. These student organizations are very proactive in providing members with scholarships, internships and career opportunities.

Outcomes

JU aviation graduates have gone on to successful careers with commercial air carriers and other aviation-related agencies. Some of their employers include COMAIR Airlines, Dallas-Fort Worth Airport, Delta Airlines, Federal Aviation Administration, Federal Express, Florida Gulf/Mesa Airlines, Jacksonville Port Authority, Naval Air Systems Command, Northrop Grumman, Philadelphia Port Authority, United Airlines, US Airways and the U.S. Navy.

B.A.S. AVIATION MAINTENANCE MANAGEMENT

A Bachelor of Applied Science in Aviation Maintenance Management programme is designed to provide a cooperative articulated Bachelor's degree for student's completing an aviation maintenance or avionics certificate/diploma programme at a technical college. The programme includes regular general education components along with aviation and management courses to enhance the academic qualifications of graduates and provide for career advancement. Forty credits will be accepted in transfer for a maintenance or avionics certificate/diploma completed at a regionally accredited post-secondary school, which may only be applied to this programme.

B.A.S. Degrees in Aviation Maintenance Management are offered by: Saint Cloud State University.

COMMERCIAL AVIATION MANAGEMENT PROGRAMME: BACHELOR OF ADMINISTRATIVE AND COMMERCIAL STUDIES (BACS)

Prepare for Take Off!

Looking for a programme that will help your future career soar to new heights? Consider the Commercial Aviation Management Programme at the University of Western Ontario, and prepare for take off!

Developed in response to the changing Canadian aviation industry, this unique programme offered by Western's Administrative and Commercial Studies Programme is the only one of its kind operated from a major Canadian university. By studying Commercial Aviation Management at Western you have the option of combining the four-year Administrative and Commercial Studies degree with formal flight training.

Graduates can look forward to rewarding careers in airport and airline management, as well as employment as pilots for private or commercial airlines.

New Aviation Environment

Today's aviation industry offers many challenging career opportunities. As air transportation systems increase in sophistication, and the role of the pilot changes, airlines are looking for employees who possess an optimal mix of academic knowledge and aviation skills.

Graduates of the Commercial Aviation Management Programme who select the flight training option will be licensed to fly commercial aircraft and be ready to assume management responsibilities, both on the ground and in the air.

COMMERCIAL AVIATION MANAGEMENT PROGRAMME

The Programme

All students in the Commercial Aviation Management stream, regardless of whether they select the flight training option, will be required to complete the following twenty full-course BACS degree programme.

Bachelor of Administrative and Commercial Studies Commercial Aviation Management

First Year

- Economics 020 (Introduction to Economics)

One full-course or equivalent from:

- Calculus 050a/b (Calculus I) and
- Calculus 051a/b (Calculus II for Math. and Physical Sciences)

Or

- Calculus 081a/b (Calculus II)
- Linear Algebra 040a/b (Linear Algebra I)
- Mathematics 028a/b (Finite Mathematics)
- Mathematics 030 (Calculus and Linear Algebra)
- Mathematics 031 (Finite Mathematics)
- Business 020 (Introduction to Business)

Or

- One designated essay full-course or equivalent numbered 020E-099E from the Faculty of Social Science or the Faculty of Arts
- Administrative and Commercial Studies 020a/b (Introduction to Administrative and Commercial Studies)

and

- Administrative and Commercial Studies 022F/G

Or

- One other half-course in Computer Science numbered 020-099
- Geography 020E (Geography: A Systems Introduction)

Or

One of:

- Physics 020 (Physics I)
- Physics 021 (Conceptual Physics for Non-Scientists)
- Physics 024 or 025 (Introductory Physics).

Second Year

- Business 257 (Accounting and Business Analysis)
- Economics 150a/b (Intermediate Micro-economic Theory I)

and

- Economics 152a/b (Intermediate Macro-economic Theory I)
- Statistical Science 135 (Elementary Statistics)

Or

- Economics 122a/b (Econometrics I)

and

- Economics 123a/b (Econometrics II)

- History 125F/G (History of Aviation)

and

- Economics 166F/G (Economics of International Air Transportation)
- Geography 115a/b (Geographic Visualization)

and

- Geography 208a/b (Weather and Climate).

Third Year

- Administrative and Commercial Studies 372 (Management Accounting)

Or

- Administrative and Commercial Studies 360a/b (Intermediate Accounting I)

and

- Administrative and Commercial Studies 361a/b (Intermediate Accounting II)
- Administrative and Commercial Studies 310a/b (Finance)

and

- Administrative and Commercial Studies 320a/b (Marketing).

One full-course or equivalent from:

- Actuarial Science 153 (Mathematics for Financial Analysis)
- Economics 154a/b (Money)
- Economics 156a/b (Labour Economics Bargaining)
- Economics 160a/b (Public Finance—Revenue)
- Economics 180a/b (Current Issues in Government Economic Policy)
- Economics 184a/b (Cost-Benefit Analysis)
- History 143F/G (Canada Since 1929)
- History 144F/G (Canadian Business and Labour)
- History 146F/G (The Cultural Mosaic and Melting Pot)
- Philosophy 162F/G (Business Ethics)
- Political Science 211E (Business and Government)
- Political Science 246E (Public Administration)
- Administrative and Commercial Studies 305a/b (Aviation Law)

and

- Administrative and Commercial Studies 306a/b (Aviation Human Factors)
- One full-course or equivalent option.

Fourth Year

- Administrative and Commercial Studies 330a/b (Operations Management)

  and

- Administrative and Commercial Studies 410b (Advanced Seminar in Administrative and Commercial Studies).

One full-course or equivalent from:

- Administrative and Commercial Studies 275a/b (Business Law)
- Administrative and Commercial Studies 325a/b
- Administrative and Commercial Studies 372 (Management Accounting) [must be completed if not taken in Third Year]
- Administrative and Commercial Studies 460a/b (Advanced Accounting I)
- Administrative and Commercial Studies 461a/b (Advanced Accounting II)
- Economics 124a/b (Economic Development I)
- Economics 162a/b (Comparative International Business)
- Economics 163a/b (International Trade)
- Economics 164a/b (International Finance)
- Geography 358F/G
- Geography 372a/b (The Geography of International Business)
- Sociology 309F/G (Demography for Business and Policy Analysis)
- One designated senior essay full-course or equivalent
- Administrative and Commercial Studies 405F/G (Airline Operation and Management).

Admission Requirements

Entry into the Commercial Aviation Management specialization is limited and highly competitive.

All students entering the Commercial Aviation Management specialization must first be accepted into the four-year Bachelor of Administrative and Commercial Studies (BACS) programme. Students may apply to enter the BACS programme when they apply for admission to the university via the Ontario Universities Application Centre.

The minimum requirement for admission to the Commercial

Aviation Management specialization of the BACS programme for Ontario students is the Ontario Secondary School Diploma (OSSD) and six Grade 12 'U' courses, including:

- Grade 12 U English (ENG4U)
- Grade 12 U Advanced Functions and Introductory Calculus (MCB4U)
- One of Geometry and Discrete Math (MGA4U) or Math of Data Management (MDM4U)
- *Note*: Grade 12 U Physics, while not required, is strongly recommended.

In addition, all students planning to receive flight training are required to possess a valid Transport Canada Category 1 Medical Certificate upon entry to the first year, and all subsequent years of the BACS programme.

It is also strongly recommended that students who wish to receive flight training have completed at least two hours of small aircraft flying time prior to making application to the Commercial Aviation Management flight training option.

It is possible that a student enrolled in the Commerical Aviation Management flight training option who has previous flight training experience may be granted appropriate credit towards the flight training component of their programme. Such credit is granted at the discretion of the Director of the BACS programme and the co-ordinator of the flight school.

Because enrolment is limited, possession of the minimum admission requirements does not guarantee admission to the BACS programme.

Bachelor of Administrative and Commercial Studies Programme,
Social Science Centre, Room 2040,
University of Western Ontario,
London, Ontario, Canada. N6A 5C2
E-mail:acs@uwo.ca
Last modified: 22 September, 2005

Tuition Fees

The University of Western Ontario tuition rate for the BACS programme is approximately $5,000* for 2005/2006.

---

* For Canadian citizens or permanent residents. Fees are subject to change.

For students entering the programme in September, 2006, the flight training portion of the programme will cost a total of $50,000. This amount is payable in four annual installments of $12,500 each, due in September of each year that the student is enrolled in the BACS programme. Therefore, students should be prepared to pay approximately $17,500 in tuition costs each year of the four-year programme ($5,000 BACS tuition + $12,500 flight training tuition).

Auxiliary fees such as the annual cost of securing the Transport Canada Category I Medical Certificate and writing the Transport Canada Private Pilot Licence, Commercial Pilot Licence, and Instrument Flight Rating examinations are not included in The University of Western Ontario tuition fee or the flight training tuition fee.

Career Opportunities

Recent market studies of the North American aviation industry are filled with encouraging news of expanding job opportunities for professionally trained pilots, especially those who have also earned a university degree.

The Commercial Aviation Management stream within Western's four-year Bachelor of Administrative and Commercial Studies (BACS) programme will equip students with an extensive set of skills, preparing them for rewarding careers in airport and airline management, as well as employment as pilots for private or commercial airlines.

All students registered in the Commercial Aviation Management stream, regardless of whether they select the flight training option, will be enrolled in the BACS degree programme. The BACS is a unique multidisciplinary programme that introduces students to the key issues in administration and commerce within the context of the social sciences. You will learn a great deal about society and social interactions, gain a capacity for critical thinking, an ability to adapt to change, and strong problem-solving and communication skills. At the same time, you will study courses in accounting, finance, marketing, and operations management. Your Commercial Aviation Management academic programme will be highlighted by senior-level courses in airline operation and management, and airport planning and management. Graduates will be well prepared to work as knowledgeable, skilled, and adaptable leaders within private and public sector business organizations generally, and within the area of airport and airline management specifically.

Commercial Aviation Management students who complete the flight training option will enjoy all of the benefits of the BACS degree, in addition to being licensed to fly commercial aircraft. Flight training will be conducted by professional flight instructors at Empire Aviation, located at London International Airport. Upon completion of their flight training, graduates will be qualified for entry-level positions as pilots in the North American aviation industry. Students will also have the option of earning a Class 4 Instructor's Rating, thereby opening the door to employment opportunities as flight instructors.

All graduates of the Commercial Aviation Management programme, flyers and non-flyers alike, will be ideally situated to benefit from expanding career opportunities in the exciting aviation industry.

For More Information

You can contact us at the following address or by filling in the online request form included below:

Bachelor of Administrative and Commercial Studies Office,
Social Science Centre, Room 2040,
University of Western Ontario,
London, Ontario, Canada. N6A 5C2
E-mail: mos@uwo.ca

This form may also be used to request additional information. Please complete ALL sections.

First Name: ....
Last Name: ....
Street Address: ....
City: ....
Province: ....
Postal Code: ....
Country: ....
E-mail Address: ....
Comments: ....

Note: If you have already made application to the University of Western Ontario to commence studies in September, 2006, and would like to be considered for admission to the Commercial Aviation Management specialization of the BACS programme, please notify the Bachelor of Administrative and Commercial Studies Office at the above address.

UNSW HANDBOOK PROGRAMME: AVIATION (MANAGEMENT)

Programme Summary

Faculty: Faculty of Science
Contact: http://www.aviation.unsw.edu.au/
Campus: Kensington Campus
Career: Undergraduate
Typical Duration: 3 years
Typical UOC per Session: 24
Min UOC per Session: 3
Max UOC per Session: 27
Min UOC for Award: 144
Award (s):
Bachelor of Aviation (Major)

Programme Description

The degree of Bachelor of Aviation is offered by the Faculty of Science with input from the Faculty of Engineering.

The aim of this programme is to provide an opportunity for students to prepare for a career in the aviation industry. Programme 3981 is a Managerial stream which prepares students more specifically for careers in the managerial sector of the aviation industry.

Both streams consists of core courses selected from the Faculties offering the programme together with a range of options. The Managerial stream offers a selection of courses designed to provide students with a broad base of knowledge in managing the operational aspects of the aviation industry.

Programme Objectives and Learning Outcomes

This programme aims to prepare students for a career in the managerial sector of the aviation industry.

Programme Structure

Stage 1

- AVEN1310 Basic Mechanics (3 UOC)

- AVEN1910 Intro to Aircraft Engineering (3 UOC)
- AVIA1321 Foundations of Aviation (6 UOC)
- AVIA1810 Aviation Security (3 UOC)
- AVIA1850 Airport Management 1 (3 UOC)
- AVIA1900 Aviation Economics (3 UOC)
- MATH1031 Mathematics for Life Sciences (6 UOC)
- MATH1041 Stats for Life & Social Sciences (6 UOC)
- PHYS1149 Physics 1A (Aviation) (6 UOC)
- SESC1580 Risk Management 1 (3 UOC)

PLUS 6 UOC from:

- ECON1101 Microeconomics 1 (6 UOC)
- PHYS1211 Energy and Environmental Physics (6 UOC)

Stage 2

- AVEN2920 Aviation Technologies 2 (3 UOC)
- AVIA1700 Aviation Safety Management 1 (3 UOC)
- AVIA2110 Aviation Human Factors 1 (3 UOC)
- AVIA2210 Aviation Human Factors 2 (3 UOC)
- AVIA2400 Aviation Regulations 1 (3 UOC)
- General Education (6 UOC)

PLUS 27 UOC from:

- AVEN2220 Aviation Eng Experimentation 1 (3 UOC)
- AVEN2910 Aviation Technologies 1 (3 UOC)
- AVEN2930 Aviation Technologies 3 (3 UOC)
- AVIA2500 Airline Marketing (3 UOC)
- AVIA2800 Management of General Aviation (3 UOC)
- AVIA3600 Simulations Applications (3 UOC)
- MGMT2721 Managing People (6 UOC)
- PHYS2810 Atmospheric Physics (3 UOC)
- SESC2580 Risk Management 2 (3 UOC)

Stage 3

- AVEN3220 Aviation Eng Experimentation 2 (3 UOC)
- AVIA3101 Airline Management (6 UOC)
- AVIA3201 Airline Resource Management (6 UOC)
- AVIA3710 Aviation Safety Management 2 (UOC)
- Further Electives totalling 6 UOC
- General Education (6 UOC)

PLUS 21 UOC from:

- AVEN3230 Aviation Systems and Avionics (3 UOC)
- AVEN3420 Aircraft Maintenance (3 UOC)
- AVEN3610 Aerodynamics, Stability & Control (3 UOC)
- AVEN3710 Aircraft Propulsion (3 UOC)
- AVEN3930 Aircraft Evaluation and Design Apprai (3 UOC)
- AVIA3300 Air Traffic Management (3 UOC)
- AVIA3400 Aviation Regulations 2 (3 UOC)
- AVIA3800 Management for Regional Airlines (3 UOC)
- AVIA3851 Airport Management 2 (6 UOC)

Stage 4 (Honours)

One of the following:

- AVIA4001 Aviation Honours—Full Time (24 UOC)
- AVIA4002 Aviation Honours—Part Time (12 UOC)

General Education Requirements

Students in this programme must satisfy the University's General Education requirements.

Academic Rules

A total of 144 units of credit including 12 units of credit of General Education (see below) are required for the completion of the Bachelor of Aviation (3 year degree).

Fees

For information regarding fees for UNSW programmes, please refer to the following website: https://my.unsw.edu.au/student/fees/FeesMainPage.html

UNSW Handbook Programme: Aviation Management

Faculty: Faculty of Science
Contact: http://www.aviation.unsw.edu.au/
Campus: Kensington Campus
Career: Postgraduate
Typical Duration: 1 year
Typical UOC per Session: 18
Min UOC per Session: 3
Max UOC per Session: 27

Min UOC for Award: 36
Award(s): Graduate Diploma (Specialisation)

Programme Description

The Graduate Diploma in Aviation Management is designed for students who have an approved diploma from a recognised tertiary institution as well as two years of relevant professional experience. Six courses will be completed to a total of 36 units of credit. A credit average must be achieved to continue on to the Masters level.

The Graduate Diploma is offered through distance education and designed with industry input for professionals and managers working in aviation-related environments. The programme can be part-time or full-time and can be completed over 2 to 6 sessions, and is further described at http://www.aviation.unsw.edu.au/.

Programme Objectives and Learning Outcomes

Please contact the Department of Aviation for information.

Programme Structure

Available Courses

- AVIA5001 Law and Regulation in Aviation (6 UOC)
- AVIA5003 Aviation and Security (6 UOC)
- AVIA5004 Av. Safety & Accident Prevention (6 UOC)
- AVIA5005 Airline Operational Management (6 UOC)
- AVIA5006 Airport Planning (6 UOC)
- AVIA5007 Airport Management (6 UOC)
- AVIA5008 Air Traffic Management (6 UOC)
- AVIA5009 Airline Corporate Management (6 UOC)
- AVIA5018 Aviation Human Factors (6 UOC)
- AVIA5019 Mgmt. of Avia. Tech. Ops. & Mainte. (6 UOC)
- AVIA5022 Aircraft Accident Invest. Tech. (6 UOC)
- AVIA5024 Flight Deck Ops. for Adv. Trans. (6 UOC)
- AVIA5311 Inflight Services Management (3 UOC)
- AVIA5312 Airline Incident Investigation (3 UOC)
- AVIA5313 Aviat'n Ground Saf'ty Invest'n (3 UOC)
- AVIA5314 Aviation System Safety (3 UOC)

Academic Rules

Please refer to the Programme Structure above and contact the school office for further information.

Fees

For information regarding fees for UNSW programmes, please refer to the following website: https://my.unsw.edu.au/student/fees/FeesMainPage.html.

UNSW HANDBOOK ON SPECIALISATION IN AVIATION

Description

The Department of Aviation in the Faculty of Science offers postgraduate studies specifically designed for the aviation industry professional who is interested in Aviation Management.

Postgraduate students can elect to enroll in the Master of Science and Technology (Aviation), the Graduate Certificate in Aviation Management or the Graduate Diploma in Aviation Management.

Studying Aviation at UNSW

The study of Aviation at UNSW is primarily through the Faculty of Science (http://www.aviation.unsw.edu.au/) with involvement from the Faculty of Engineering. Several postgraduate programmes are offered.

Jetstar Graduate Opportunities

Jetstar Airways are seeking graduates with a passion for the aviation industry who are looking for the opportunity to build a meaningful career with Jetstar.

The Programme

The Jetstar Airways Graduate Programme offers candidates the choice of three streams: Corporate, Commercial and Operations. We strongly encourage candidates to apply for the stream that is of most interest to them.

The programme will run for 12 months in duration and upon successful completion graduates will be offered a permanent position, subject to suitable vacancies company wide and satisfactory performance over the year.

Expected Outcomes

It is expected that Jetstar Graduates will gain invaluable industry-specific work experience that will provide them with the tools and exposure to build their knowledge, skills and working relationships to forge a successful career in aviation.

For more information regarding the Jetstar Airways Graduate Programme and to apply, please visit our website "Jetstar.com", select "about us" from the top toolbar, "work with us" from the drop down menu and click "Jetstar Airways Graduate Programme 2007."

Applications for Entry in 2007: UNSW Aviation

Applications are currently being accepted for entry into both undergraduate and postgraduate programmes in Aviation at UNSW.

Undergraduate Applications

Australian Local and Australian Year 12 International Students are to submit their applications via the Universities Admissions Centre (UAC) for both the Management and Flying streams.

All other international students who hold overseas qualifications must apply to the Direct Admissions office. More information can be found from the UNSW International Office website.

In addition to the formal application, flying stream applicants are required to submit an internal application to the UNSW Aviation department. An interview will be scheduled upon receiving this internal application. The internal application form is available from the UNSW Aviation office.

As part of the application process for the Flying Stream, applicants are required to complete a Class 1 medical examination prior to enrolment. The medical examination is required to be completed by a Designated Aviation Medical Examiners (DAME). A full list of all medical examiners is available on the CASA website.

Before applicants can present for their medical examination, they are required to obtain an Aviation Reference Number (ARN). This ARN will become Pilot licence number and reference for all aspects of your aviation career. To obtain your ARN, please complete the following steps:

1. Complete the ARN application form (available from UNSW Aviation or any CASA office)—Form 1162.
2. Attach a copy of one form of identification showing your full name and date of birth.

3. Submit the completed application and form of identification to your nearest CASA office.
4. Allow two (2) weeks for processing.
5. When you receive your ARN, take it with you when you present for your medical examination.

Note: Management stream applicants are not required to attend the interview or complete the aviation medical examination.

Postgraduate Applications

Applications for entry into the postgraduate coursework programmes by distance education can be made directly to the University via the online application website.

There are three entry points all having varying entry requirements:

1. Graduate Certificate in Aviation Management (programme code 7448).
2. Graduate Diploma in Aviation Management (programme code 5678).
3. Master of Science and Technology (Aviation) (programme code 8738).

Once the online application has been submitted, applicants are to send certified copies of their supporting documentation (curriculum vitae, license, logbook, academic results, and diplomas) to the Direct Admissions Office.

Direct Admissions Office
The University of New South Wales
Sydney, NSW 2052
Australia.

GLOBAL AIR TRAINING: A CASE STUDY

Global Air Training is a specialist training company, providing training, development and consultancy services in Aviation and Healthcare in every continent of the world. Based in the Aviation industry, we provide a range of training programmes to meet the needs of airlines, military and other aviation operators. Programmes include Crew Resource Management, Human Factors, Safety and Security. Our Aviation training programmes are recognised and accredited by regulatory bodies worldwide including the UK Civil Aviation Authority and Ministry of Defence.

In Healthcare, our Team Resource Management based training programmes are leading the development of patient safety focused systems in healthcare organisations. Clients include leading international airlines worldwide, government regulatory bodies, the UK National Health Service (NHS) and private healthcare organisations.

Aviation Training Services

Global Air Training (GAT) is a specialist aircrew training provider. The company was founded in 1996. Clients now include over 300 international airlines, civil aviation authorities, military, V.I.P. and head of state aviation operations based in every continent of the world. GAT provides initial, recurrent and train the trainer programmes for the aviation industry tailored to the specific needs and culture of the client organisation and the operating regulatory environment.

Training programmes available from GAT include:

- Crew Resource Management (CRM)
- Crew Resource Management Instructor (CRMI) CAA approved core course
- Maintenance Resource Management
- Maintenance Human Factors Instructor
- Safety and Emergency Procedures (SEP)
- Aircrew Survival
- Managing Disruptive Passenger Behaviour
- Aviation Security
- Aviation First-Aid
- Automated External Defibrillation (AED)
- Cabin Service
- Conflict Management

Ministry of Defence

Global Air Training is an approved provider (Ref.1713) for UK Ministry of Defence Enhanced Learning Credits (ELC).

Crew Resource Management

Crew Resource Management (CRM) studies the application of human factors knowledge to enhance the safety and efficiency of flight operations. Global Air Training provides training and consultancy in Crew Resource Management for over 200 airlines worldwide. We provide customised CRM training, CRM programme design and Train-the-Trainer programmes. Our *Crew Resource Management Instructor (CRMI)* course is recognised worldwide and approved by the UK CAA as a core course for JAR CRMI accreditation. We also operate a unique *Cabin Crew CRMI* course.

Maintenance Human Factors

The European Aviation Safety Authority (EASA) requires that personnel involved in any maintenance, management and/or quality audits, must have training in human factors and human performance capabilities.

GAT Maintenance Human Factors courses are fully compliant with the human factors training requirements of EASA in Part 145.

The Maintenance Human Factors (MHF) Train the Trainer course is appropriate for those personnel who are involved with the delivery of MHF training to licensed aircraft engineers and all personnel working in the aircraft maintenance environment.

Crew Resource Management (CRM)

Global Air Training provides a full range of training and development services in support of airline Crew Resource Management and Human Factors training. We have more than 10 years experience in developing and delivering CRM programmes for airlines and a diverse range of military and government aviation organisations. Clients include over 250 airlines and leading aviation operations based in every continent of the world. Services available include CRM programme design, CRM courseware design, research, direct initial and recurrent CRM training for crew, training and development for ground, line and simulator instructors. All GAT CRM facilitators are experienced commercial and/or military aircrew and are accredited by the UK CAA. In addition to 10 years experience of airline flight deck CRM training, we have extensive experience of CRM training in the cabin environment, corporate aviation sector and for single crew civil and military operations. We offer a unique *Cabin Crew Resource Management Instructor* training course.

CRM Initial/Recurrent Training

Successful CRM training must always take into account of the individual nature and challenges of the client organisation. Global

Air Training works with each client to provide effective, customised CRM training programmes to meet the individual needs and culture of the organisation.

CRM Programme Design

We can provide a complete design service for modular CRM programme design and high quality courseware design for delivery by an operators internal CRM facilitators.

Our CRM courseware and case studies are extensive and constantly revised to ensure that they are operator relevant, current and reflect the latest developments and best practice.

Instructor/Facilitator Development

We offer a full range of CRM instructor courses including the UK CAA approved CRMI core course, and assessment and accreditation services for ground, line and simulator instructors.

Crew Resource Management Instructor (CRMI)

The CAA approved core course is suitable for delegates wishing to attain CRMI accreditation for ground or line training. The programme content meets the standard CRM requirements and recommendations of: JAR, CAA, FAA and ICAO. The 'Train the Trainer' course is 5 days in duration, covering theoretical aspects of modern CRM training with relevant references to, and examination of human factor related aviation incidents and accidents. The course utilises video footage, CVR recordings and investigation board reports.

The course includes instructional techniques, facilitation, critique and feedback techniques, NOTECHS and assessment. Following the course the student will have the knowledge, foundation and materials to deliver a professional CRM course suitable for the pilots, or the combined pilots and cabin crew of Airline operations. The course includes:

- 5 days classroom training,
- CRM Instructor Training Manual in hardcopy and on CD,
- Full set of course handouts both on paper and CD-ROM,
- PowerPoint presentation on CD,
- Copyright permission for reproduction and utilisation of course manuals and handouts for use within your airline,
- Certification on completion of course, and

- Investigation reports and studies are available for delegates to use during the CRM course.

Courses at GAT UK also include:

- Courtesy transport to and from Manchester (MAN) or Liverpool (LPL) Airports.
- Daily transport between local hotel and our training centre.
- Refreshments and lunch on course training days.

Cabin Crew Resource Management Instructor (CRMI)

The Cabin CRMI course is designed for Cabin Crew CRM instructors who will be involved in Cabin Crew CRM training to meet the requirements of JAR-OPS 1 Subpart 0. The 5-day course covers all aspects of current CRM training in the context of the cabin environment with relevant references to, and examination of human factor related aviation incidents and accidents. The course utilises cabin based case studies, video footage and investigation board reports. The course includes instructional techniques, facilitation, briefing, debriefing and feedback techniques. Following the course the student will have the knowledge, skills and materials to deliver professional CRM training suitable for cabin crew in airline operations, and meeting the requirements of JAR-OPS 1 Subpart 0.

Human Factors in Aviation Maintenance, MHF Instructor Programme

A core course for instructors who will be involved in the development and delivery of training in Human Factors in Aviation Maintenance (MHF).

EASA 145 Requirements

EASA part 145 requires that personnel involved in any maintenance, management and/or quality audits, must have an understanding of the application of human factors and human performance capabilities. Part 145 approved maintenance organisations are required to have completed the initial training by 28th September, 2006. This course will be of particular interest to organisations required to meet the Human Factors training requirements of EASA 145.

Maintenance Human Factors (MHF)

The aim of MHF training is to increase safety, quality and efficiency in aircraft maintenance operations by reducing human error and its impact on maintenance activities.

The course syllabus fully covers the MHF training requirements for EASA 145 compliant engineering organisations. The 'Train the Trainer' course is 5 days in duration; covering theoretical aspects of MHF training with relevant references to, and examination of human factor related maintenance incidents and accidents.

The course utilises video footage, case studies and investigation board reports. The course develops practical skills including instructional techniques, facilitation, critique and feedback.

Throughout the course, we promote an awareness of our behavioural attitudes to improve and develop communication and interpersonal relationships with all those involved in the aviation system. The training addresses the nature of your company's operations and procedures, which produce particular difficulties and unusual hazards.

Course Objectives

Following the course the student will:

- have the knowledge, skills and materials to introduce, deliver and maintain an effective EASA 145 compliant MHF training programme within their organisation.
- be able to deliver appropriate MHF training for licensed engineers, managers and all personnel working in the aircraft maintenance environment.

The Course Includes

- 5 days classroom training.
- Comprehensive MHF Instructor Training Manual in hardcopy and on CD.
- Full set of course handouts both on paper and CD-ROM.
- PowerPoint presentation on CD.
- Copyright permission for reproduction and utilisation of course manuals and handouts for use within your organisation.
- Certification on completion of course.

- Investigation reports and studies.

The MHF Instructor programme is typically 5 days in duration. However, the content and duration of the programme may be adjusted to meet local needs.

Contemporary Human Factors endeavours have been targeted towards flight deck and cabin crew performance with Crew Resource Management training being compulsory for crews for more than a decade. Additionally, albeit to a lesser extent, Human Factors training has also been applied in air traffic control. Most available literature shows less concern for the Human Factors issues which affect aircraft maintenance personnel who inspect, diagnose and repair aircraft. However, statistically, human error in aircraft maintenance has had a significant effect upon flight safety. The implementation of a Human Factors programme for aircraft maintenance engineers is probably now one of the key areas where improvements in safety can be made.

EASA 145 Requirements

JAR 145 regulations are being superseded by the European Aviation Safety Authority (EASA) part 145 regulations. EASA requires that personnel involved in any maintenance, management and/or quality audits, must have an understanding of the application of human factors and human performance capabilities. Part 145 approved maintenance organisations are required to have completed the initial training by 28th September, 2006.

Maintenance Human Factors Training

Maintenance Human Factors (MHF) training examines the importance of human factors to aircraft maintenance engineering. It is essential to all personnel operating in an aircraft maintenance and engineering environment.

The aim of MHF training is to increase safety, quality and efficiency in aircraft maintenance operations by reducing human error and its impact on maintenance activities.

GAT Maintenance Human Factors Course

The course fulfils the MHF training requirements for EASA 145 approved engineering organisations. The course covers theoretical aspects of MHF training with relevant references to, and examination of human factor related maintenance incidents and accidents. The course utilises video footage, case studies and investigation board reports.

Throughout the course, we promote an awareness of our behavioural attitudes to improve and develop communication and interpersonal relationships with all those involved in the aviation system. The training addresses the nature of your company's operations and procedures, which produce particular difficulties and unusual hazards.

The course uses both classroom instruction and practical exercises including group discussions and incident reviews.

Objectives

The objectives are:

- To enhance maintenance personnel's awareness of individual and organisational human factors issues, both positive and negative, that they may affect airworthiness.
- To develop MHF skills, in order to make a positive impact on the safety and efficiency of maintenance operations.

Safety and Emergency Procedures Courses

The objective of SEP training courses is to provide the crew member with the knowledge, skills and confidence to deal effectively with in-flight emergency situations, and thereby to maximise the probability of an optimum and safe outcome for crew and passengers. The objectives are realised through a combination of lectures, case studies, discussions and practical hands on training. Each training course is specific to the client's aircraft fleet, safety equipment, nature of operations and standard operating procedures.

SEP Courses

Global Air Training provides the following SEP courses, at our training facility, and at operators bases, worldwide:

- Pilot SEP,
- Cabin crew initial,
- Cabin Crew Recurrent,
- SEP Instructors—Train the Trainer, and
- Bespoke—To meet operator requirements.

Security Key Courses

Global Air Training provides the following Security courses, at our training facility, and at operators bases, worldwide:

- Aviation Security,
- Aviation Security Managers and Instructors,
- Managing Disruptive Passenger Behaviour,
- Security for Freight Forwarder, and
- Bespoke—To meet operator requirements.

Aviation Security

Our security training courses reflect the current climate and heightened threat levels. Objectives are to:

- increase awareness of the current threat to aviation,
- to enable the recognition of incendiary and explosive devices and their component parts,
- to understand and apply the correct precautions to reduce the threat of interference to aviation and to gain knowledge of the procedures that should be adopted in the event of a breach of aircraft security, and
- GAT instructors are qualified to UK government standards and experienced in both airline and airport security operations.

Worldwide Training Services

Global Air Training hosts aviation security training courses at our modern and well-equipped training facility in the United Kingdom, close to Manchester International Airport.

We can also assist in the development of bespoke courses for your organisation and present training via field instructors at client bases worldwide.

Managing Disruptive Passenger Behaviour

Situational Control and Physical Intervention Techniques for Aircrew

In response to cases of disruptive passenger behaviour, Global Air Training has developed this new programme, with input from airlines and experts in disruptive passenger handling in other industries. Research indicates:

- The largest proportion of all in-flight incidents will require only verbal intervention if dealt with using correct conflict management techniques.
- If an incident becomes physically threatening, many of the currently taught defensive techniques are inappropriate and ineffective under pressure given only a few hours of training.
- There will often be a large physical strength disparity between cabin crew and a disruptive, aggressive passenger
- Disruptive passenger handling procedures must be developed or updated in the context of a secure cockpit door policy.
- The current security environment must be considered.

Our objective is to deliver a programme which can be taught in one-day. The delegate will acquire skills in detecting, preventing and eliminating potential incidents of disruptive, unruly or criminal passenger behaviour, and effectively managing such incidents when they do occur whilst maintaining personal and passenger safety.

In one Application, this Training System Resulted in an 80 per cent Decrease in Attacks on Staff

The delegate will develop techniques and principles which are easily remembered and used successfully, even when under pressure, following only one-day of training.

Delegates learn to:

- understand the Corporate policy, State and International law on disruptive passengers.
- recognise and categorise a potentially disruptive situation
- recognise behavioural markers that indicate escalation of the situation.
- select and apply appropriate intervention techniques to de-escalate the situation. .
- maintain personal and passenger safety.
- become proficient in the techniques which may be used to control a disruptive passenger.
- effectively use self-defence techniques in a violent or life-threatening situation.

- understand the procedure of handing over a restrained passenger to the appropriate authorities.

Outcome

"By the end of the session all candidates will be able to recognise and defuse verbally potentially violent encounters and apply physical intervention techniques should the need arise; this will be tested under pressure."

Aviation First Aid

The objectives of aviation first aid training are to enable the crew member to assess a situation and condition by observing the signs, symptoms and where possible by eliciting any applicable medical history of the casualty. To make a diagnosis on the information they have collated, to administer appropriate treatment immediately and to maintain treatment of the casualty until professional medical help takes over. To give the crew member the confidence and ability to preserve life, prevent a condition worsening and to promote recovery. The objectives are realised through a combination of lectures, case studies, discussions and practical hands on training.

Automated External Defibrillation, AED

A critical key to saving victims of sudden cardiac arrest is early access to defibrillation. Chances of survival decrease by 7 to 10 per cent for each minute that passes without defibrillation. A victim's best chance for survival is when there is revival within 4 minutes.

An Automated External Defibrillator (AED) is a small, lightweight device used to assess a person's heart rhythm. If necessary, it administers an electric shock to restore a normal rhythm in victims of sudden cardiac arrest.

Provision of AED equipment allows crew members with proper training to defibrillate cardiac arrest victims quickly and effectively, improving the chances of survival dramatically.

There is no legal bar to airlines making a defibrillator available to air crews if the assessment of first aid needs indicates such equipment may be required. The Health and Safety (First-Aid) Regulations 1981, do not prevent someone who is specially trained from taking action beyond the initial management of a casualty. It is essential that the person who will be required to use a defibrillator is appropriately trained.

This course covers the use of automated external defibrillation equipment.

Course Content

- Revision of CPR,
- Anatomy and Physiology of the heart,
- Improving survival chances using AED,
- Undertaking AED, and
- Selecting and maintaining AED equipment.

BCAS TRAINING IN INDIA: A STUDY OF PATTERN

Delhi

1. Basic course at Delhi for Air Sahara from 11-22 December, 2006.
2. Basic course at Delhi for Airlines from 04-15 December, 2006.
3. Basic course at Delhi, Gurgaon for SpiceJet from 27 November-06—December, 2006.
4. Basic course at Delhi for Air India from 30th October-10 November, 2006.
5. Basic course at Delhi for IAL from 02-12 December, 2006.
6. Basic course at Pathankot for Punjab Police and Airdeccan from 13-23 November, 2006.
7. Basic course at Delhi for SpiceJet, Gurgaon from 19-29 September, 2006.
8. Basic course at Delhi for Airlines from 31 July to 11 August, 2006.
9. Basic course at Cargo Service Centre Delhi from 31 July to 10 August, 2006.
10. Basic course at SKPA Udhampur, Delhi, from 26 June to 06 July 2006.
11. Basic course at Delhi, for CISF from 20-30 June, 2006.
12. Basic course at Delhi, for CISF from 06-16 June, 2006.
13. Basic course at Delhi, Gurgaon for SpiceJet from 20-30 June, 2006.
14. Basic course at Delhi for CISF from 09-19 May, 2006.
15. Basic course at Delhi for Airlines from 15-25 May, 2006.

16. Basic course at Delhi for Indigo from 22 May to 01 June, 2006.
17. Basic course at Delhi for CISF from 24 April to 05 May, 2006.
18. Basic course at Delhi for CISF from 05-15 April, 2006.
19. Basic course at Delhi for Indus Airlines from 05-15 April, 2006.
20. Basic course at Delhi for CISF from 21 Febuary-03 March, 2006.
21. Basic course at (HQ BCAS) Delhi for SpiceJet from 03-13 April, 2006.
22. Basic course at (HQ BCAS) Delhi for Airlines from 13-24 March, 2006.
23. Basic course at Delhi for CISF from 7-17 March, 2006.
24. Basic course at Delhi for Airlines from 28 February to 10 March, 2006.
25. Basic course at Delhi for CISF from 31 January to 10 February, 2006.
26. Basic course at Delhi for Air Sahara from 13-23 February, 2006.
27. Basic course at Delhi for Airlines from 06-17 February, 2006.
28. Basic course at Delhi for CISF from 09-20 January, 2006.
29. Basic course at Delhi for IAL from 02-12 January, 2006.

Mumbai

1. Basic course at Mumbai for Jet Airways from 02-13 February, 2007.
2. Basic course at Mumbai for Kingfisher from 06-16 February, 2007.
3. Basic course at Mumbai for Airlines from 09-19 January, 2007.
4. Basic course at Mumbai for Jet Airways from 04-15 January, 2007.
5. Basic course at Mumbai for Air India from 11-22 December, 2006.
6. Basic course at Mumbai for Jet Airways from 09-20 November, 2006.
7. Basic course at Mumbai for Security Personnel from 13 November to 24 November, 2006.

8. Basic course at Mumbai for CISF from 06 November to 17 November, 2006.
9. Basic course at Mumbai for Air India from 14 November to 24 November, 2006.
10. Basic course at Mumbai for Go Air from 25 October to 04 November, 2006.
11. Basic course at Mumbai for Kingfisher from 17-27 October, 2006.
12. Basic course at Mumbai for Jet Airways from 06-16 October, 2006.
13. Basic course at Mumbai for CISF from 03-13 October, 2006.
14. Basic course at Mumbai from 14-25 September for Jet Airways 2006.
15. Basic course at Kandla Mumbai for Gujarat Police from 20-30 September, 2006.
16. Basic course at Mumbai for CISF from 18-29 September, 2006.
17. Basic course at Mumbai for CISF from 05-15 September, 2006.
18. Basic course at Mumbai for Go Air from 21-31 August, 2006.
19. Basic course at Mumbai for CISF from 23 August to 01 September, 2006.
20. Basic course at Mumbai for Jet Airways from 10-21 August, 2006.
21. Basic course at Mumbai for Airlines from 4-15 September, 2006.
22. Basic course at Mumbai for Jet Airways from 13-24 July, 2006.
23. Basic course at Mumbai for CISF from 24 July to 04 August, 2006.
24. Basic course at Mumbai for Air India from 31 July to 11 August, 2006.
25. Basic course at Mumbai for CISF 09-20 January, 2006.
26. Basic course at Mumbai for CISF 19-30 June, 2006.
27. Basic course at Mumbai for Jet Airways 15-26 June, 2006.
28. Basic course at Mumbai for Air India from 29 May to 09 June, 2006.
29. Basic course at Mumbai for CISF from 23 January, to 03 February, 2006.

30. Basic course at Mumbai for CISF from 29 May to 09 June, 2006.
31. Basic course at Mumbai for Kingfisher from 16-26 May, 2006.
32. Basic course at Mumbai for Airlines from 02-12 May, 2006.
33. Basic course at Mumbai for CISF from 8-19 May, 2006.
34. Basic course at Mumbai for Kingfisher from 03-13 May, 2006.
35. Basic course at Mumbai for CISF from 17-28 April, 2006.
36. Basic course at Mumbai for CISF from 20-31 March, 2006.
37. Basic course at Mumbai for Jet Airways from 16-31 March, 2006.
38. Basic course at Mumbai for Jet Airways from 09-24 March, 2006.
39. Basic course at Mumbai for Go Air from 13-23 March, 2006.
40. Basic course at Mumbai for Air India from 08-21 March, 2006.
41. Basic course at Mumbai for CISF from 06-18 March, 2006.
42. Basic course at Mumbai for CISF from 13-24 February, 2006.
43. Basic course at Mumbai for Go Air from 16-27 January, 2006.
44. Basic course at Mumbai for Kingfisher Airlines from 09-19 January, 2006.
45. Basic course at Mumbai for Airlines from 16-27 January, 2006.
46. Basic course at Mumbai for Air India from 09-19 January, 2006.
47. Basic course at Mumbai for Jet Airways from 09-20 January, 2006.

Chennai

1. Basic avsec course from 22 January-02 February, 2007, for Air India at Chennai.
2. Basic avsec course from 05-17 February, 2007, for CISF NISA at Hyderabad Chennai.
3. Basic avsec course from 04-15 December, November, 2006, for CISF NISA at Hyderabad Chennai.
4. Basic avsec course from 11-22 December, 2006, for Air Deccan at Bangalore.
5. Basic avsec course from 11-21 November, 2006, for

Paramount and Blue Dart at Chennai.

6. Basic avsec course from 13-23 November, 2006, for CISF NISA at Hyderabad Chennai.
7. Basic avsec course from 13-23 November, 2006, for CISF 17th Batch NISA at Hyderabad Chennai.
8. Basic avsec course from 13-23 November, 2006, for IAL at Hyderabad Chennai.
9. Basic avsec course from 07-17 November, 2006, for Airdeccan at Bangalore.
10. Basic avsec course from 16-28 October, 2006, for Airlines at Chennai.
11. Basic avsec course from 26th October, 07 November, 2006, for Karnataka Police at Vidyanagar.
12. Basic avsec course from 09-19 October, 2006, for IAL Hyderabad-Chennai.
13. Basic avsec course from 03-13 October, 2006, for Air Deccan at Bangalore-Chennai.
14. Basic avsec course from 28 August-07 September, 2006, for Air India at Chennai.
15. Basic avsec course from 21-31 August, 2006, for CISF NISA at Hyderabad Chennai.
16. Basic avsec course from 5-15th September, 2006, for Air Deccan at Bangalore.
17. Basic avsec course from 21-31 August, 2006, for IAL at Hyderabad Chennai.
18. Basic avsec course from 17-27 July, 2006, for CISF NISA at Hyderabad Chennai.
19. Basic avsec course from 5-15th July, 2006, for Indigo at Bangalore.
20. Basic avsec course from 03-13 July, 2006, for Air Deccan at Bangalore.
21. Basic avsec course from 13-24 July, 2006, for Jet Airways at Chennai.
22. Basic avsec course from 19-29 June, 2006, for IAL, CTE at Hyderabad.
23. Basic avsec course from 12-22 June, 2006, for CISF at NISA Hyderabad.
24. Basic avsec course from 19-30 June, 2006, for Airlines RDCOS at Chennai.
25. Basic avsec course from 15-25 May, 2006, IAL CTE at Hyderabad.

26. Basic avsec course from 09-19 May, 2006, Air Deccan Bangalore.
27. Basic avsec course from 27 March, to 07 April, 2006 at Hyderabad.
28. Basic avsec course from 06-17 March, 2006at Chennai.
29. Basic avsec course from 27 March-07 April, 2006 for Air India at Cochin.
30. Avsec course from 9-20 April, 2006 for IAL at Chennai.
31. Basic avsec course from 16-27 March, 2006 for CISF at Chennai.
32. Basic avsec course from 6-17 March, 2006, for Air Deccan at Bangalore.
33. Basic avsec course from 02-12 January, 2006, for CISF at Chennai.
34. Basic avsec course from 10-20 January, 2006, for Air Deccan at Bangalore.
35. Basic avsec course from 23 January to 03 February, 2006 for NISA, CISF Hyderabad.
36. Basic avsec course from 16-27 January, 2006, for IAL personnel at Chennai.
37. Basic avsec course from 09-20 January, 2006 at NISA Hyderabad.
38. Basic avsec course from 18-30 January, 2006 for CISF personnel at Chennai.
39. Basic avsec course from 26 December, 2005 to 05 January 2006 at Hyderabad.

Kolkata

1. Basic course at Kolkata from 11-21 November, 2006, for Indigo Air.
2. Basic course at Kolkata from 11-21, September, 2006 for Air India.
3. Basic course at Kolkata from 03-14 July, 2006 for Airlines.
4. Basic course at Guwahati Kolkata for Indigo from 21 June to 02 July, 2006.
5. Basic course at Kolkata from 17-28 April, 2006 for Airlines.
6. Basic course at Kolkata for CISF from 04-14 April, 2006.
7. Basic course at Kolkata for IAL from 16-27 March, 2006.
8. Basic course at Kolkata for CISF from 14-24 March 2006.
9. Basic course at Kolkata for CISF from 21 February to 03 March, 2006.

10. Basic course at Kolkata for CISF from 06-16 February, 2006.
11. Basic course at Kolkata for CISF from 09-20 January, 2006.

AVSEC Testing and Certification of X-ray Screeners Course

Delhi

1. X-ray screeners from 15-16 February, 2007 at RDCOS Delhi.
2. X-ray screeners from 04-05 January, 2007 at RDCOS Delhi.
3. X-ray screeners from 08-09 January, 2007 at RDCOS Delhi.
4. X-ray screeners from 26-27 December, 2006 at RDCOS Delhi.
5. X-ray screeners from 18-19 December, 2006 at RDCOS Delhi.
6. X-ray screeners from 07-08 December, 2006 at RDCOS Delhi.
7. X-ray screeners from 28-29 December, 2006 at RDCOS Delhi.
8. X-ray screeners from 22-23 November, 2006 at RDCOS Delhi.
9. X-ray screeners from 18-19 October, 2006 at RDCOS Delhi.
10. X-ray screeners from 15-16 November, 2006 at RDCOS Delhi.
11. X-ray screeners from 07-08 November, 2006 at RDCOS Delhi.
12. X-ray screeners from 26-27 October, 2006 at RDCOS Delhi.
13. X-ray screeners from 27-28 September, 2006 at RDCOS Delhi.
14. X-ray screeners from 09-10 October, 2006 at RDCOS Delhi.
15. X-ray screeners from 06-07 September, 2006 at RDCOS Delhi.
16. X-ray screeners from 25-26 July, 2006 at RDCOS Delhi.
17. X-ray screeners from 27 -28 July, 2006 at RDCOS Delhi.
18. X-ray screeners from 06-07 July, 2006 at RDCOS Delhi.
19. X-ray screeners from 29-30 June, 2006 at RDCOS Delhi.
20. X-ray screeners from 14-15 June, 2006 at RDCOS Delhi.
21. X-ray screeners from 06-07 June, 2006 at RDCOS Delhi.
22. X-ray screeners from 29-30 May, 2006 at RDCOS Delhi.
23. X-ray screeners from 03-04 May, 2006 at RDCOS Delhi.
24. X-ray screeners from 26-27 April, 2006 at RDCOS Delhi.
25. X-ray screeners from 27-28 March, 2006 at RDCOS Delhi.
26. X-ray screeners from 5-7 April, 2006 at RDCOS Delhi.
27. X-ray screeners from 7-8 March, 2006 at RDCOS Delhi.
28. X-ray screeners from 27-28 February, 2006 at RDCOS Delhi.
29. X-ray screeners from 01-02 February, 2006 at RDCOS Delhi.

30. X-ray screeners from 12-13 January, 2006 at RDCOS Delhi.
31. X-ray screeners from 30-31 January, 2006 at RDCCS Delhi.
32. X-ray screeners from 05-06 January, 2006 at RDCOS Delhi.

Kolkata Region

1. X-ray screeners from 12-13 February, 2007 at rdCOS Kolkata.
2. X-ray screeners from 18-19 January, 2007 at RDCOS Kolkata.
3. X-ray screeners from 04-05 January, 2007 at RDCOS Kolkata.
4. X-ray screeners from 18-19 December, 2006 at RDCOS Kolkata.
5. X-ray screeners from 13-14 December, 2006 at RDCOS Kolkata.
6. X-ray screeners from 05-06 December, 2006 at RDCOS Kolkata.
7. X-ray screeners from 27-28 November, 2006 at RDCOS Kolkata.
8. X-ray screeners from 21-22 November, 2006 at RDCOS Kolkata.
9. X-ray screeners from 13-14 November, 2006 at RDCOS Kolkata.
10. X-ray screeners from 06-07 November, 2006 at RDCOS Kolkata.
11. X-ray screeners from 26-27 October, 2006 at RDCOS Kolkata.
12. X-ray screeners from 18-19 October, 2006 at RDCOS Kolkata.
13. X-ray screeners from 09-10 October, 2006 at RDCOS Kolkata.
14. X-ray screeners from 20-21 September, 2006 at RDCOS Kolkata.
15. X-ray screeners from 20-21 September, 2006 at RDCOS Kolkata.
16. X-ray screeners from 04-05 September, 2006 at RDCOS Kolkata.
17. X-ray screeners from 24-25 July, 2006 at RDCOS Kolkata.
18. X-ray screeners from 26-27 June, 2006 at RDCOS Kolkata.
19. X-ray screeners from 29-30 May, 2006 at RDCOS Kolkata.
20. X-ray screeners from 07-09 June, 2006 at RDCOS Kolkata.
21. X-ray screeners from 08-09 May, 2006 at RDCOS Kolkata.
22. X-ray screeners from 12-13 April, 2006 at RDCOS Kolkata.
23. X-ray screeners from 03-04 April, 2006 at RDCOS Kolkata.

24. X-ray screeners from 13-14 March, 2006 at RDCOS Kolkata.
25. X-ray screeners from 21-22 March, 2006 at RDCOS Kolkata.
26. X-ray screeners from 22-23 February, 2006 at RDCOS Kolkata.
27. X-ray screeners from 09-10 February, 2006 at RDCOS Kolkata.
28. X-ray screeners from 19-20 January, 2006 at RDCOS Kolkata.
29. X-ray screeners from 03-04 January, 2006 at RDCOS Kolkata.

Mumbai

1. X-ray screeners from 29-31 January, 2007 at RDCOS Mumbai.
2. X-ray screeners from 19-20 February, 2007 at RDCOS Mumbai.
3. X-ray screeners from 14-15 February, 2007 at RDCOS Mumbai.
4. X-ray screeners from 10-11 February, 2007 at RDCOS Mumbai.
5. X-ray screeners from 07-08 February, 2007 at RDCOS Mumbai.
6. X-ray screeners from 15-16 January, 2007 at RDCOS Mumbai.
7. X-ray screeners from 04-05 January, 2007 at RDCOS Mumbai.
8. X-ray screeners from 28-29 December, 2006 at RDCOS Mumbai.
9. X-ray screeners from 21-22 December, 2006 at RDCOS Mumbai.
10. X-ray screeners from 18-19 December, 2006 at RDCOS Mumbai.
11. X-ray screeners from 29-30 November, 2006 at RDCOS Mumbai.
12. X-ray screeners from 04-05 December, 2006 at RDCOS Mumbai.
13. X-ray screeners from 27-28 November, 2006 at RDCOS Mumbai.
14. X-ray screeners from 15-16 November, 2006 at RDCOS Mumbai.
15. X-ray screeners from 20-21 November, 2006 at RDCOS Mumbai.
16. X-ray screeners from 09-10 November, 2006 at RDCOS Mumbai.
17. X-ray screeners from 23-24 October, 2006 at RDCOS Mumbai.

18. X-ray screeners from 11-12 October, 2006 at RDCOS Mumbai.
19. X-ray screeners from 05-06 October, 2006 at RDCOS Mumbai.
20. X-ray screeners from 26-27 September, 2006 at RDCOS Mumbai.
21. X-ray screeners from 18-19 September, 2006 at RDCOS Mumbai.
22. X-ray screeners from 11-12 September, 2006 at RDCOS Mumbai.
23. X-ray screeners from 28-29 August, 2006 at RDCOS Mumbai.
24. X-ray screeners from 17-18 August, 2006 at RDCOS Mumbai.
25. X-ray screeners from 15-16 May, 2006 at RDCOS Mumbai.
26. X-ray screeners from 31 July to 01 August, 2006 at RdCOS Mumbai.
27. X-ray screeners from 11-12 July, 2006 at RDCOS Mumbai.
28. X-ray screeners from 28-29 June, 2006 at RDCOS Mumbai.
29. X-ray screeners from 13-14 June, 2006 at RDCOS Mumbai.
30. X-ray screeners from 26-27 May, 2006 at RDCOS Mumbai.
31. X-ray screeners from 18-19 April, 2006 at RDCOS Mumbai.
32. X-ray screeners from 05-06 April, 2006 at RDCOS Mumbai.
33. X-ray screeners from 16-17 February, 2006 at RDCOS Mumbai.
34. X-ray screeners from 09-10 March, 2006 at RDCOS Mumbai.
35. X-ray screeners from 16-17 March, 2006 at RDCOS Mumbai.
36. X-ray screeners from 30-31 January, 2006 at RDCOS Mumbai.
37. X-ray screeners from 07-08 February, 2006 at RDCOS Mumbai.
38. X-ray screeners from 05-06 January, 2006 at RDCOS Mumbai

Chennai

1. X-ray screeners from 5-6 February, 2007 at RDCOS Chennai.
2. X-ray screeners from 24-25 January, 2007 at RDCOS Chennai.
3. X-ray screeners from 10-11 January, 2007 at RDCOS Chennai.

4. X-ray screeners from 28-29 December, 2006 at RDCOS Chennai.
5. X-ray screeners from 13-14 December, 2006 at RDCOS Chennai.
6. X-ray screeners from 21-22 November, 2006 at RDCOS Chennai.
7. X-ray screeners from 29-30 November, 2006 at RDCOS Chennai.
8. X-ray screeners from 02-03 November, 2006 at RDCOS Chennai.
9. X-ray screeners from 25-26 October, 2006 at RDCOS Chennai.
10. X-ray screeners from 05-06 October, 2006 at RDCOS Chennai.
11. X-ray screeners from 30-31 October, 2006 at RDCOS Chennai.
12. X-ray screeners from 11-12 October, 2006 at RDCOS Chennai.
13. X-ray screeners from 07-08 August, 2006 at RDCOS Chennai.
14. X-ray screeners from 06-07 September, 2006 at RDCOS Chennai.
15. X-ray screeners from 21-22 September, 2006 at RDCOS Chennai.
16. X-ray screeners from 20-21 July, 2006 at RDCOS Chennai.
17. X-ray screeners from 10-11 July, 2006 at KOCHI Chennai.
18. X-ray screeners from 10-11 May, 2006 at RDCOS Chennai.
19. X-ray screeners from 29-30 May, 2006 at RDCOS Chennai.
20. X-ray screeners from 13-14 June, 2006 at RDCOS Chennai.
21. X-ray screeners from 20-21 March, 2006 for Air India at Kozhikode.
22. X-ray screeners from 20-21 April, 2006 at RDCOS Chennai.
23. X-ray screeners from 12-13 April, 2006 at RDCOS Chennai.
24. X-ray screeners from 28-29 March, 2006 at RDCOS Chennai.
25. X-ray screeners from 02-03 March, 2006 at RDCOS Chennai.
26. X-ray screeners from 13-14 February, 2006 at RDCOS Chennai.
27. X-ray screeners from 06-07 February, 2006 at RDCOS Chennai.

28. X-ray screeners from 21-22 February, 2006 at RDCOS Chennai.
29. X-ray screeners from 17-18 January, 2006 at RDCOS Chennai.
30. X-ray screeners from 24-25 January, 2006 at RDCOS Chennai.

AVSEC National Auditor's Course

Delhi

- AVSEC National auditors course from 25 April- 03 May, 2006, BCAS Headquaters.

AVSEC Supervisors Course

Delhi

- AVSEC supervisors course from 07-11 August, 2006 at BCAS Headquaters.

AVSEC Cabin Crew Course

1. AVSEC Cabin Crew course at MDLR Airlines at Gurgaon from 24 January-02 February, 2007.
2. AVSEC Cabin Crew course at Mumbai for Go Air from 04-09 December, 2006.
3. AVSEC Cabin Crew course at Mumbai for Go Air from 20-25 November, 2006.
4. AVSEC Cabin Crew course at Mumbai for Go Air from 25-31 October, 2006.
5. AVSEC Cabin Crew course for Indigo Gurgaon from 24-31 August, 2006.
6. AVSEC Cabin Crew course for SpiceJet Gurgaon from 07-12 August, 2006.
7. AVSEC Cabin Crew course for Indigo at NIAMAR from 21-28 July, 2006.
8. AVSEC Cabin Crew course at Mumbai for Go Air from 03-08 July, 2006.
9. AVSEC Cabin Crew course for Jagson Airlines from 13-28 July, 2006.
10. AVSEC Cabin Crew course for Indigo at NIAMAR from 22-28 June, 2006.
11. AVSEC Cabin Crew course for SpiceJet from 14-20 June, 2006.

12. AVSEC Cabin Crew course for SpiceJet Gurgaon from 08-13 April, 2006.
13. AVSEC Cabin Crew course for Indus Air from 10-17 March, 2006.

AVSEC Management Course

Delhi

- AVSEC Management course for Airlines from 27 Febuary-09 March, 2006

AVSEC Instructor's Course

Delhi

1. AVSEC Instructor's course at BCAS Hqrs. from 23 October-01 November, 2006.
2. AVSEC Instructor's course at BCAS Hqrs. from 05 -13 June, 2006.
3. AVSEC Instructor's course at BCAS Hqrs. from 23-31 January, 2006.

AVSEC Cargo Course

Delhi

1. AVSEC cargo course at BCAS Hqrs. from 25-29 September, 2006.
2. AVSEC cargo course at BCAS Hqrs. from 27-31 March, 2006.

GLOSSARY OF AVIATION EDUCATION RELATED TERMS

Advanced Standing

Advanced Standing, sometimes referred to as credit transfer, at undergraduate level is recognition of prior study at a University or TAFE. At postgraduate level it is recognition of prior postgraduate study in a university.

Award

An award is a degree, diploma or certificate obtained when a student graduates from a programme at UNSW. It recognises the student's successful completion of that programme.

Bachelor

A Bachelor degree is the formal award a student receives when they successfully complete an undergraduate university degree programme, ordinarily of three or more years duration.

Campus

This is the teaching location where a programme, course or plan is taught. UNSW has several campuses including the main campus at Kensington, the College of Fine Arts campus in Paddington, and the Australian Defence Force Academy in Canberra.

Co-major

A co-major is part of a sequence of study for a program in which the requirements for two majors are met.

Combined Programme

A combined programme is a programme of study which leads to the award of two degrees, that is, the graduate earns two qualifications (an example of this would be the Bachelor of Arts/Bachelor of Laws). These are also sometimes called combined degrees. They have a single set of programme rules.

Core Course

A course which is compulsory and must be passed in order to complete the requirements of the programme.

Corequisite

A corequisite is a course which must be completed successfully before, or studied concurrently with, the course for which it is prescribed.

Course

Otherwise known as a subject, a course is an individual study unit offered within a programme and plan (for example, MATH1131—Mathematics 1A). Students enroll in many courses to make-up their programme of study, some of which may be core courses (courses which need to be completed to satisfy the requirements of a particular programme) and some of which may be elective courses (where students are given a choice of courses). At UNSW, courses are identified by a four character alphabetic prefix which identifies the subject area or specialisation administering the course and a four-digit numeric suffix e.g. ECON1101 Microeconomics 1.

Coursework

Coursework refers to a mode of study which is largely or wholly constituted of courses involving face-to-face class instruction. It is a term which is commonly used with regard to undergraduate and postgraduate study. The other mode of postgraduate study is research.

Degree

A degree is the formal qualification awarded when a student graduates from an undergraduate programme of study such as a Bachelor of Arts, or a postgraduate Masters or Ph.D programme.

Doctorate/Doctoral Programme

A doctoral programme is a postgraduate research programme where students independently research a specific topic under the guidance of a supervisor to produce a thesis. For a doctorate, considerably more original work is required than for a Masters by Research programme. Students should note that in some faculties, coursework may also be prescribed.

Exclusion

Exclusions are courses students are excluded from taking, generally because they have content in common with courses for which the student has previously been granted credit.

Faculty

Faculties are the large academic organisational units of the University, and are generally comprised of several schools or departments. UNSW has eight faculties. Arts and Social Sciences; Built Environment; Commerce and Economics; Engineering; Law; Medicine; Science; and the College of Fine Arts. University College, ADFA and the Australian Graduate School of Management are also regarded as faculties.

Fast-track Programme

UNSW offers several "fast-track" or "Masters track" programmes. These give students the opportunity, if they meet progression requirements, to progress directly from an undergraduate programme to a particular Masters programme with some courses in the final year counting towards both qualifications e.g. the Bachelor of Engineering/Master of Commerce.

General Education

UNSW requires undergraduate students to complete some courses outside the study area of the degree programme in which they are enrolled. General Education courses are offered in a variety of general subject areas to allow students to complete this requirement.

International Student

International students are citizens of a country other than Australia or New Zealand and are not Australian permanent residents.

Local Student

Local students are Australian citizens, Australian permanent residents or New Zealand citizens.

Major

Many programmes require students to complete a major. A major is an approved sequence of study in an area of academic or vocational specialisation. This is also sometimes referred to as 'a'.

Masters

A Masters programme or degree is a postgraduate programme where students enrol in an approved sequence of courses involving face-to-face instruction. Some Masters programmes also involve a research component.

Minor

In some programmes, students are required to supplement their study major (see above) with a 'minor.' This is a sequence of study in a secondary area of specialisation, comprising fewer units of credit than a major (usually 24). For example, a student enrolled in an Science degree programme might complete a major in Anatomy and a minor in Zoology. Majors and minors are both examples of a 'plan'.

Non-Award

Non-award enrolment means that the course/s undertaken by the student do not lead to the award of any formal degree, diploma or certificate at UNSW. Students from other universities ("cross institutional") often enrol in non-award courses at UNSW, as credit may be granted for these courses by their home institution.

Plan

A plan is a sequence of study within a programme focused on a particular study area, usually requiring students to complete an approved sequence of 'core' and 'elective' courses. At UNSW, plans are identified by a five-digit alphabetical prefix and a five-digit numeric suffix e.g. SENGA13648 refers to the full-time Software Engineering plan.

Postgraduate

Postgraduate programmes of study are available to students who have already completed a university degree programme in a related area. They offer the opportunity for students to further their skills and qualifications in a particular area of specialisation. Completion of a postgraduate programme may lead to an award of a Graduate Certificate, Graduate Diploma, Masters (by Coursework or Research), Doctorate (Ph.D) or post-doctoral qualification.

Prerequisite

Some courses have prerequisites. A pre-requisite is a requirement which must be completed before enrolling in the course or the next level of courses e.g. completing a Level I MATH course before progressing to Level II MATH courses.

Programme

A programme is an approved programme of study which leads to the award of a degree, diploma or certificate. Programmes may be undergraduate or postgraduate and are identified by a four-digit numeric code e.g. the programme code for the Bachelor of Psychology is 3432.

Research

Research programmes of study are postgraduate programmes of study which involve a student independently researching a specific topic under the guidance of a supervisor and producing a thesis or report. Some research programmes do involve a coursework component.

School

This is an academic organisational unit, also sometimes referred to as a department. Faculties may be comprised of several schools e.g. the Faculty of Arts and Social Sciences has 12 schools including the School of Philosophy and the School of History.

Session/Semester

A session or semester is a university teaching period. Each academic year, there are two main sessions (Sessions 1 and 2), usually of 14 weeks teaching, plus an examination period. There are also shorter Summer and Winter sessions that run during the breaks between the major sessions. Exceptions to this pattern are the Faculty of Medicine and the Australian Graduate School of Management whose academic years are divided into four teaching periods.

Specialisation

A specialisation is an area of academic expertise on which students focus their studies, often by enrolling in a plan offered in that area, such as a Philosophy major within an Arts degree. Examples of specialisations include French, Biological Science, Taxation etc.

Stage

Programmes are generally structured in a number of 'stages' of study, requiring students to complete a specified number of units of credit and/or a particular sequence of courses at each stage. Generally, when a student completes their degree programme within the normal minimum time, the different stages will correspond with the different years of the student's enrolment (e.g. Level 1 is Year 1, Level 2 is Year 2, etc).

Undergraduate

Undergraduate programmes of study are degree programmes which do not require students to have previously undertaken university study in order to enrol. They are designed for students who have completed secondary studies (high school) in Australia or have a level of education deemed equivalent to this (e.g., equivalent overseas study or alternate entry programmes).

Unit of Credit

Each course at UNSW has a particular load or weighting which is referred to as a unit of credit e.g. the course ELEC1101 Electrical Engineering is worth 3 units of credit. This is often abbreviated to UOC. UNSW programmes require the successful completion of a certain number of UOCs and fees are also charged on a UOC basis.

4

Aviation Management: Frequently Asked Questions

AVIATION MANAGEMENT

Q. Who is the Degree Coordinator for Aviation Management?

Ans. The Degree Coordinator for Aviation Management is Dr. Gerald Fairbairn. His phone number is 603-577-6456; his e-mail is fairbairn@dwc.edu.

Q. What types of jobs are available for me after I graduate with an aviation management degree?

Ans. With an aviation management degree quite a few jobs are available to you. You can become an airline manager, usually starting out as a lead, then working your way-up to the management department within a few years. You have the capability and knowledge required to start your own commuter operation, FBO, or even aircraft sales. There are also a wide variety of jobs within the FAA available to you and within air traffic control departments.

Q. What types of courses will I be taking in DWC's aviation management programme?

Ans. The DWC courses you will be taking are outlined in DWC's catalogue. They consist of business courses, aviation management courses, airport management courses, aircraft sales, and humanities, plus financial accounting and marketing.

Q. Can I take any flight operation courses while enrolled in DWC's aviation management programme?

Ans. You are eligible to take elective courses for credits, in which you may take flight operation courses. However, DWC's flight programme is often at capacity with flight operation student majors, so it may typically be difficult for aviation management students to enroll in flight labs.

Q. What are the starting salaries for aviation management majors?

Ans. Starting salaries vary for DWC aviation management students after graduation. A rough estimate would be around $25,000 with the potential to earn an average of $130,000-$150,000, depending on the kind of job and location.

Q. Are there internships available for aviation management students?

Ans. Internships are readily available for DWC students. Typical internships include Angel Flight, FAA, Hoyle Tanner and Associates, Gale Associates, and Pease International Authority.

Q. How much does it cost to enroll into the DWC aviation management programme? Are there any extra lab costs?

Ans. All aviation management courses are included in the standard tuition cost at DWC. There are no extra lab costs even though we have field trips to provide students with first hand experience. These include trips to Logan International Airport, Washington DC, the FAA center and other facilities.

Q. Are the professors at DWC experienced in the aviation industry?

Ans. DWC professors are all trained with current information involving the aviation industry. All of our professors have recently worked or have been involved in the aviation industry.

Q. What makes the aviation management programme at DWC better than taking any other business course at another college?

Ans. DWC's aviation management programme gives you hands-on knowledge as to what to expect in the aviation industry, whereas other colleges typically give you required business and management skills, but not the aviation background. DWC requires a series of business courses along with management courses that will enable you to become a manager not only in aviation, but also in virtually any other industry.

Q. Why do some students switch their major to aviation management after being in the flight programme?

Ans. Some DWC students switch their majors to Aviation Management because of the cost of the flight programme. DWC's flight programme is very challenging and dedication is required for success in the programme. DWC's aviation management degree is an excellent alternative.

METROPOLITAN STATE UNIVERSITY PROGRAMME ON AVIATION MANAGEMENT

Q. What are the admission requirements for this programme?

Ans. If you are transferring into the university with at least 16 post-secondary semester credits at a grade point average of at least 2.00, no entrance exams are required as a prerequisite for admission to Metropolitan State University. If you have less than 16 semester credits, you will be admitted as a freshman if your standardized college admission test score (ACT, PSAT or SAT) is at or above the national average, or you graduated from high school in the top 50 per cent of your class. Most students will be required to take assessment tests in reading, writing and mathematics to determine placement in appropriate classes. For more information please check the Admissions Office website or call 651-793-1306.

Q. How do I apply for admission to the programme?

Ans. To apply for admission, you must complete a Metropolitan State University application and check off "Aviation Management." Send the completed application form, along with a check for $20.00 and photocopies of both sides of your FAA and FCC licenses, to the Metropolitan State University Admissions Office, 700 E. 7th St., St. Paul, MN 55106-5000. You must also make a written request to all other schools you have attended to send all transcripts directly to the Admissions Office.

Q. How many credits will transfer into this programme from other schools?

Ans. If you attended another university or community college with regional accreditation, all credits earned with a grade of at least C—will transfer to Metropolitan State University. The extent to which those credits will satisfy your graduation requirements will depend on the individual courses. If you graduated from one of the following Minnesota technical college programme, you will receive 40 semester credits, which can be transferred only into this programme:

- Minneapolis
  Aviation Technology (A&P)
- Thief River Falls
  Aviation Technology (A&P)
- Winona
  Aviation Technology (A&P)
- Alexandria
  Avionics
- Hutchinson
  Non-Destructive Testing
- Dakota County
  Technical Management
- Lake Superior
  Technical Management.

General education courses taken at technical colleges may transfer as well. Other credits will be evaluated on an individual basis.

Q. Where and when will the classes be held?

Ans. Most aviation classes are held at various locations around the Twin Cities area. Some courses are online. Most of the general education classes meet on Metropolitan State University's campuses in downtown (MCTC) Minneapolis, the Midway area in St. Paul, and east St. Paul campus. Campus maps are included online and in the class schedule.

Q. How will my FAA and FCC certificates be evaluated for transfer?

Ans. You can be awarded up to 20 semester credits for your FAA and FCC certificates. These credits will be included as general electives and will count toward the total of 128 semester credits required for graduation in this programme.

Q. Can I get credit for prior learning?

Ans. Aviation-related work prior learning experiences can be evaluated for upto 12 semester credits total, 1-4 variable credits per experience, if the experience is determined to have sufficient academic and practical content, will be added to your transcript and counted among your aviation electives. Tuition will be charged for credits earned in this manner just as if you had taken a traditional class. Examples of college-level learning that may be appropriate for prior learning credits might include ground or flight instruction, unlicensed dispatch, airport management.

Q. How many classes can I take at once?

Ans. There is no specific limit on the number of courses you can take during a single semester; however, students who also have full-time jobs or who have not attended school for a significant period of time are generally advised to take no more than two courses at a time.

Q. What courses are required for graduation?

Ans. A total of 128 semester credits are required for graduation. Of those credits, 40 must be taken in specified general education subjects, including communications, mathematics, science, social and behavioural science, humanities, computers, cultural and global awareness, and 8 upper-division credits in liberal studies. 23-28 credits must be taken in the Aviation Management core, and 12-18 credits are required in aviation electives. You must take a total of at least 40 upper-division credits. Students satisfy the residency requirement by taking at least 30 semester credits through the consortium of universities offering the Aviation Management programme.

Q. How will I know which classes to take?

Ans. You will receive direction and assistance in deciding which courses to take during the Perspectives class, which you should take during your first or second semester. The Programme Advisor can provide additional guidance. Students who enter the university with less than 16 semester credits must also take the Your Academic Journey, Metro 101 course, and most students must take assessment tests for placement in appropriate reading, writing and mathematics classes.

Q. Will independent study courses be offered?

Ans. Independent study courses are included among the aviation electives; these must be approved by faculty from the designated university. Students may also take many other courses through Metropolitan State as faculty-designed or student-designed independent studies.

Q. How long will it take to complete the degree?

Ans. The exact amount of time needed to complete the degree will be different for each individual. It will depend on a number of factors, including the number of credits initially transferred into the programme, the number of courses taken each semester, and the timing and scheduling of the required and elective courses.

Q. Where can I buy textbooks?

Ans. Textbooks may be purchased at the Metropolitan State Bookstore on the St. Paul Campus, 651-793-1670.

Q. Is there an orientation process for new students?

Ans. All entering students must attend a new student orientation session before or during their first semester. During these sessions, which are held on selected evenings and Saturdays, students will be provided with information about Metro State's programmes, facilities, services, registration procedures and financial aid.

Q. How much will the classes cost?

Ans. Each of the four participating schools determines its own tuition rates for the courses it offers; these rates are likely to change from year to year. You may obtain current tuition information from each university's web site:
Metropolitan State University.
Minnesota State University, Mankato.
St. Cloud State University.
Winona State University.

Q. How can I apply for financial aid?

Ans. Information about student loans, VA benefits, grants and scholarships can be obtained by calling the financial aid office at Metropolitan State University at 651-793-1414, or on the Financial Aid Office website.

Q. What arrangements must I make to pay tuition?

Ans. If you are not receiving financial aid, you must pay each university separately for its courses. Metropolitan State requires payment at the time of registration, either by check or credit card. St. Cloud State University and Minnesota State University-Mankato also accept credit card payments.

Q. Can I take classes during the summer?

Ans. Students may take independent study courses in aviation with the permission of the instructor. Metropolitan State also offers many elective and general education classes during the summer.

Q. Will daytime classes be offered?

Ans. At this time all of the aviation courses are being offered during the evening, Monday through Thursday. Daytime classes may be offered in the future if there is sufficient demand.

Q. Are job placement services available?

Ans. Metropolitan State University provides a number of services for students, including a Career Services Office, 651-793-1528

AVIATION TECHNOLOGY—PURDUE UNIVERSITY

Q. What career opportunities are open to someone with a degree in this programme?

Ans. Students with this degree normally seek careers in airline operations, which includes flight dispatch, station management, flight schedule coordination, air cargo administration, public relations, and marketing or customer service. Many graduates pursue careers in airport management, and some seek positions within state or government aviation agencies.

Q. I want to combine management and flight, is that possible with this major?

Ans. Yes, each student must choose an area of concentration, and flight is one of the choices. To complete a flight specialty a student must complete the private pilot certificate and the commercial and instrument rating through Purdue or outside.

Q. I want to transfer into the Flight Programme.

Ans. If your goal is to transfer into the Flight programme at Purdue, realize that the competition for transfer is fierce. There are only two times during your academic career when you will have an opportunity to transfer: (1) at the conclusion of your first semester, (2) at the end of your sophomore year. Transfer students will be selected based upon the number of vacant positions and upon your relative college grade point average. In general an A-to a B+ grade point average must be achieved. But in certain circumstances, such as few vacancies and many transfer requests, an even higher grade point average may be required. In addition, students who attempt to transfer after their first semester must have already attained the FAA Private Pilot Certificate. Students transferring at the conclusion of their sophomore year must have received their Commercial and Instrument ratings.

Q. May I take flight courses at Purdue?

Ans. If space is available, it may be possible to enroll in flight courses offered at Purdue without being in the Flight programme. Space is generally more available during the summer semester and sometimes in the spring semester. Students may receive flight training at any number of organizations including, but not limited to the Purdue Pilots, or Lafayette Aviation.

Q. Is it difficult to transfer into a different programme if I decide this is not the major for me?

Ans. It depends on which programme you are interested in transferring into. Make sure you discuss this with your advisor before you register for classes.

Q. Is there a residence hall where most aviation students live?

Ans. No, aviation students live in many different residence halls. As you consider different ones, you may want to think about if you prefer co-ed halls, high-rise or low-rise, close to main campus, close to the airport, close to the recreational gym, or close to Mackey Arena.

Q. Will my parentsÕ insurance cover my belongings in a residence hall or apartment?

Ans. Check with your parents about what coverage you will have on items you plan to bring to campus.

Q. May I bring my car to campus?

Ans. A car is usually not necessary because the Aviation Technology Department does run a free bus from main campus to the airport to get you back and forth to campus anytime classes are in session. There is also a city bus system. However, if you feel you need your car, contact your residence hall about a parking permit or Visitor Information if you are living in an apartment.

Q. With this major should I bring a computer?

Ans. It is not required that you purchase or have your own computer however it is helpful. There are many computer laboratories on campus and students are welcome to use the computer lab at the airport during certain hours. Remember that you will be required to write papers for various classes and if you are dependent on computer labs you must plan ahead.

Q. Can I be involved in college sports with this major?

Ans. Yes, we have many students involved in college sports including swimming, track, basketball and others. Students plan their schedule with their academic advisor and student athletes are able to complete the plan of study within the four years.

Q. What activities are available to aviation students?

Ans. Purdue offers many clubs and organizations depending on your interests. This is a time to experiment with what sounds interesting. You can find activities to sports, band and music, politics, journalism, arts and others. Activities are ways to make friends, relive stress, build a resume, develop leadership skills and have fun!

Q. Are many aviation students involved in ROTC?

Ans. There are aviation students involved in all branches of the ROTC programmes. Different ROTC programmes have different requirements so you will need to talk with the recruiter to know what your requirements would be.

Q. What is Day on Campus?

Ans. Day on Campus is a one day programme in June designed to give students an orientation to their department, register for classes, take advance placement exams, visit your assigned residence hall, and have financial aid questions answered.

Q. Is it necessary to attend Day on Campus in the summer?

Ans. Everyone is encouraged to attend Day on Campus but we do realize that for some students it is not possible. If this is the case for you, contact the department and make special arrangements.

Q. May I take flight courses at Purdue?

Ans. If space is available, it may be possible to enroll in flight courses offered at Purdue without being in the Flight programme. Space is generally more available during the summer semester and sometimes in the spring semester. Students may receive flight training at any number of organizations including, but not limited to, the Purdue Pilots or Lafayette Aviation.

Q. What is Gold Rush?

Ans. Gold Rush is a student orientation held before classes begin and designed for freshmen to help them adjust to campus. There are many activities to introduce you to other freshmen and campus services.

Department of Aviation Technology

Purdue University
1401 Aviation Drive
West Lafayette IN 47907-2015
Voice: (765) 494-5782
Fax: (765) 494-2305

College of Technology

Knoy Hall
401 N.Grant St.
West Lafayette, IN 47907
For School information, e-mail: info@tech.purdue.edu
For Website issues, contact the COT Webmaster.

AERONAUTICAL TECHNOLOGY PROGRAMME

Q. What career opportunities are open to someone with a degree in this programme?

Ans. This programme prepares students for career positions in aircraft design, manufacturing, modification and operation. Students are being recruited for positions such as project planning, airline maintenance management, project management, technical publication, operations planning, customer support and engineering liaison.

Q. If I am also interested in flying, can I combine that with this programme?

Ans. Yes, each student must choose an area of concentration and flight is one of the choices. To complete a flight specialty a student must complete the private pilot certificate and the commercial and instrument rating through Purdue or outside.

Q. What is the FAA Airframe and Power Plant (A&P) Mechanic certificate?

Ans. The Federal Aviation Administration (FAA) regulates all aviation operations in the United States. The FAA must certificate any person performing or supervising aircraft for return to service after maintenance. The certificate is recognized throughout the aviation industry as evidence that the holder has acquired a high level of technical knowledge and skill.

Q. When will I need to buy tools?

Ans. Students are required to purchase a basic tool set by the beginning of your first semester.

Q. How this is major different from Freshman Engineering?

Ans. Students in Aeronautical Technology are introduced to aviation courses the first semester. Most courses are a combination of lecture and lab, which benefits the student who learns by hands on experience. Engineering requires students to take higher levels of math and science before they begin their engineering area.

Q. Is it difficult to transfer into a different programme if I decide this is not the major for me?

Ans. It depends on which programme you are interested in transferring into. Make sure you discuss this with your advisor before you register for classes.

Q. Is there a residence hall where most aviation students live?

Ans. No, aviation students live in many different residence halls. As you consider different ones, you may want to think about if you prefer co-ed halls, high-rise or low-rise, close to main campus, close to the airport, close to the recreational gym, or close to Mackey Arena.

Q. Will my parents insurance cover my belongings in a residence hall or apartment?

Ans. Check with your parents about what coverage you will have on items you plan to bring to campus.

Q. May I bring my car to campus?

Ans. A car is usually not necessary because the Aviation Technology Department does run a free bus from main campus to the airport to get you back and forth to campus anytime classes are in session. There is also a city bus system. However, if you feel you need your car, contact your residence hall about a parking permit or Visitor Information if you are living in an apartment.

Q. With this major should I bring a computer?

Ans. It is not required that you purchase or have your own computer however it is helpful. There are many computer laboratories on campus and students are welcome to use the computer lab at the airport during certain hours. Remember that you will be required to write papers for various classes and if you are dependent on computer labs you must plan ahead.

Q. Can I be involved in college sports with this major?

Ans. Yes, we have many students involved in college sports including swimming, track, basketball and others. Students plan their schedule with their academic advisor and student athletes are able to complete the plan of study within the four years.

Q. What activities are available to aviation students?

Ans. Purdue offers many clubs and organizations depending on your interests. This is a time to experiment with what sounds interesting. You can find activities connected to sports, band and music, politics, journalism, arts and others. Activities are ways to make friends, relive stress, build a resume, develop leadership skills, and have fun!

Q. Are many aviation students involved in ROTC?

Ans. There are aviation students involved in all branches of the ROTC programmes. Different ROTC programmes have different requirements so you will need to talk with the recruiter to know what your requirements would be.

Q. What is Day on Campus?

Ans. Day on Campus is a one day programme in June designed to give students an orientation to their department, register for classes, take advance placement exams, visit your assigned residence hall, and have financial aid questions answered.

Q. Is it necessary to attend Day on Campus in the summer?

Ans. Everyone is encouraged to attend Day on Campus but we do realize that for some students it is not possible. If this is the case for you, contact the department and make special arrangements.

Q. May I take flight courses at Purdue?

Ans. If space is available, it may be possible to enroll in flight courses offered at Purdue without being in the Flight programme. Space is generally more available during the summer semester and sometimes in the spring semester. Students may receive flight training at any number of organizations including, but not limited to, the Purdue Pilots or Lafayette Aviation.

Q. What is Gold Rush?

Ans. Gold Rush is a student orientation held before classes begin and designed for freshmen to help them adjust to campus. There are many activities to introduce you to other freshmen and campus services.

Department of Aviation Technology

Purdue University
1401 Aviation Drive
West Lafayette IN 47907-2015
Voice: (765) 494-5782
Fax: (765) 494-2305

College of Technology

Knoy Hall
401 N.Grant St.
West Lafayette, IN 47907
For School information, e-mail: info@tech.purdue.edu
For Website issues, contact the COT Webmaster

FLIGHT PROGRAMME

Q. Is 20/20 uncorrected vision necessary to pass the medical? (What is a medical?)

Ans. No, students may have corrected vision with the exception of being a military pilot. To pursue a career in flight your vision should be correctable to 20/20. Students must possess a second-class medical certificate when classes begin. For a list of doctors that perform the exam, check with your physician or local airport.

Q. How many flight hours will I have when I finish?

Ans. The number of hours vary depending on how much flying beyond the requirements a student has pursued. Many students become flight instructors and are able to increase their hours by working summers as a flight instructor. The average number of hours at graduation ranges from 300 to 500.

Q. What are the career opportunities with this degree programme when I finish?

Ans. The programme trains students to be pilots either for an airline or for a corporation. However, most students will not have enough flight time to be considered by the airlines and will spend time flight instructing to build their hours.

Q. How long does it take after graduation to be employed by the major airlines?

Ans. It is very hard to say how long it will take for you because it always depends on the market. In the past, it has taken between two and seven years after graduation for a significant number of students to be hired as pilots for a major airline.

Q. Is it possible to double major in flight and something else?

Ans. It depends on which programme you are interested in transferring into. Make sure you discuss this with your advisor before you register for classes.

Q. Is it difficult to transfer into a different programme if I decide this is not the major for me?

Ans. It depends on which programme you are interested in transferring into. Make sure you discuss this with your advisor before you register for classes.

Q. Is there a residence hall where most aviation students live?

Ans. No, aviation students live in many different residence halls. As you consider different ones, you may want to think about if you prefer co-ed halls, high-rise or low-rise, close to main campus, close to the airport, close to the recreational gym, or close to Mackey Arena.

Q. Will my parentsÕ insurance cover my belongings in a residence hall or apartment?

Ans. Check with your parents about what coverage you will have on items you plan to bring to campus.

Q. May I bring my car to campus?

Ans. A car is usually not necessary because the Aviation Technology Department does run a free bus from main campus to the airport to get you back and forth to campus anytime classes are in session. There is also a city bus system. However, if you feel you need your car, contact your residence hall about a parking permit or Visitor Information if you are living in an apartment.

Q. With this major should I bring a computer?

Ans. It is not required that you purchase or have your own computer however it is helpful. There are many computer laboratories on campus and students are welcome to use the computer lab at the airport during certain hours. Remember that you will be required to write papers for various classes and if you are dependent on computer labs you must plan ahead.

Q. Can I be involved in college sports with this major?

Ans. Yes, we have many students involved in college sports including swimming, track, basketball and others. Students plan their schedule with their academic advisor and student athletes are able to complete the plan of study within the four years.

Q. What activities are available to aviation students?

Ans. Purdue offers many clubs and organizations depending on your interests. This is a time to experiment with what sounds interesting. You can find activities connected to sports, band and music, politics, journalism, arts and others. Activities are ways to make friends, relive stress, build a resume, develop leadership skills, and have fun!

Q. Are many aviation students involved in ROTC?

Ans. There are aviation students involved in all branches of the ROTC programmes. Different ROTC programmes have different requirements so you will need to talk with the recruiter to know what your requirements would be.

Q. What is Day on Campus?

Ans. Day on Campus is a one day programme in June designed to give students an orientation to their department, register for classes, take advance placement exams, visit your assigned residence hall, and have financial aid questions answered.

Q. Is it necessary to attend Day on Campus in the summer?

Ans. Everyone is encouraged to attend Day on Campus but we do realize that for some students it is not possible. If this is the case for you, contact the department and make special arrangements.

Q. May I take flight courses at Purdue?

Ans. If space is available, it may be possible to enroll in flight courses offered at Purdue without being in the Flight programme. Space is generally more available during the summer semester and sometimes in the spring semester. Students may receive flight training at any number of organizations including, but not limited to, the Purdue Pilots or Lafayette Aviation.

Q. What is Gold Rush?

Ans. Gold Rush is a student orientation held before classes begin and designed for freshmen to help them adjust to campus. There are many activities to introduce you to other freshmen and campus services.

Department of Aviation Technology

Purdue University
1401 Aviation Drive
West Lafayette IN 47907-2015
Voice: (765) 494-5782
Fax: (765) 494-2305

College of Technology

Knoy Hall
401 N.Grant St.
West Lafayette, IN 47907
For School information, e-mail: info@tech.purdue.edu
For Website issues, contact the COT Webmaster.

AERONAUTICAL/FLIGHT COMBINED PROGRAMME

Q. What ratings and certificates will I have when I finish this programme?

Ans. You will have a private pilot certificate, commercial, instrument and multi-engine ratings and your airframe and power plant certificate.

Q. When will my flight classes begin?

Ans. You will be enrolled in the ground school for your private pilot certificate the first semester and will begin flying the second semester.

Q. What are the career opportunities for students in this programme?

Ans. This programme prepares students for career positions in aircraft design, manufacturing, modification and operation. It also includes all the flight ratings that are needed to be a professional pilot.

Q. When will I need to buy tools?

Ans. Students are required to purchase a basic tool set by the beginning of your first semester.

Q. Is there additional financial aid available for this programme?

Ans. Yes, the Division of Financial Aid will consider the additional costs associated with flight classes and tool costs when determining your financial aid package.

Q. Is it difficult to transfer into a different programme if I decide this is not the major for me?

Ans. It depends on which programme you are interested in transferring into. Make sure you discuss this with your advisor before you register for classes.

Q. Is there a residence hall where most aviation students live?

Ans. No, aviation students live in many different residence halls. As you consider different ones, you may want to think about if you prefer co-ed halls, high-rise or low-rise, close to main campus, close to the airport, close to the recreational gym, or close to Mackey Arena.

Q. Will my parents insurance cover my belongings in a residence hall or apartment?

Ans. Check with your parents about what coverage you will have on items you plan to bring to campus.

Q. May I bring my car to campus?

Ans. A car is usually not necessary because the Aviation Technology Department does run a free bus from main campus to the airport to get you back and forth to campus anytime classes are in session. There is also a city bus system. However, if you feel you need your car, contact your residence hall about a parking permit or Visitor Information if you are living in an apartment.

Q. With this major should I bring a computer?

Ans. It is not required that you purchase or have your own computer however it is helpful. There are many computer laboratories on Campus and students are welcome to use the computer lab at the airport during certain hours. Remember that you will be required to write papers for various classes and if you are dependent on computer labs you must plan ahead.

Q. Can I be involved in college sports with this major?

Ans. Yes, we have many students involved in college sports including swimming, track, basketball and others. Students plan their schedule with their academic advisor and student athletes are able to complete the plan of study within the four years.

Q. What activities are available to aviation students?

Ans. Purdue offers many clubs and organizations depending on your interests. This is a time to experiment with what sounds interesting. You can find activities connected to sports, band and music, politics, journalism, arts and others. Activities are ways to make friends, relive stress, build a resume, develop leadership skills, and have fun!

Q. Are many aviation students involved in ROTC?

Ans. There are aviation students involved in all branches of the ROTC programmes. Different ROTC programmes have different requirements so you will need to talk with the recruiter to know what your requirements would be.

Q. What is Day-on-Campus?

Ans. Day-on-Campus is a one-day programme in June designed to give students an orientation to their department, register for classes, take advance placement exams, visit your assigned residence hall, and have financial aid questions answered.

Q. Is it necessary to attend Day-on-Campus in the summer?

Ans. Everyone is encouraged to attend Day-on-Campus but we do realize that for some students it is not possible. If this is the case for you, contact the department and make special arrangements.

Q. May I take flight courses at Purdue?

Ans. If space is available, it may be possible to enroll in flight courses offered at Purdue without being in the Flight programme. Space is generally more available during the summer semester and sometimes in the spring semester. Students may receive flight training at any number of organizations including, but not limited to, the Purdue Pilots or Lafayette Aviation.

Q. What is Gold Rush?

Ans. Gold Rush is a student orientation held before classes begin and designed for freshmen to help them adjust to campus. There are many activities to introduce you to other freshmen and campus services.

CANADIAN AVIATION INSTITUTE

Q. What are the mark requirements for the programme?

Ans. Applicants should have a minimum of 60 per cent overall in both Grade 12 maths and Grade 12 English (general level).

Q. What are the admission requirements for the programme?

Ans. Specific admission requirements are listed on the Admission Requirements by Programme page.

Q. What are the admission requirements for a student coming from out of province?

Ans. Information for admission requirements for out-of-province students is found on the Admissions Information page, under the heading "Studied Outside of Ontario or Canada?"

Q. When can I start the programme?

Ans. The only time that you can start the CAvI is in September.

Q. How do I apply?

Ans. You must apply online through OCAS (Ontario College Application Service).

Q. I am currently attending another college/university—can I come to the CAvI?

Ans. Exemptions will be offered depending on the college/university programme that you are taking or have completed. The details for exemptions are dealt with on an individual basis. The students who come to the CAvI with previous education are considered "Fast-Trackers" and may be able to complete the programme in less than 3 years.

Q. What is the deadline for applications?

Ans. Application deadlines are listed on the OCAS (Ontario College Application Service) website.

Q. How much is tuition?

Ans. Tuition fees are listed on the Fees and Web Registration page.

Q. How do I get a tour of the Canadian Aviation Institute?

Ans. If you would like to come and visit the CAvI for a tour, please contact (705) 722-5129 to set-up a date and time. The tours can be flexible to your schedule.

Q. Can I sit in on some classes?

Ans. If you would like to come to the CAvI to sit in on a class, you are more than welcome to come! Please e-mail aviation@georgianc.on.ca.

Q. Is there financial aid available?

Ans. Information on the types of financial aid available to students is found on the Financial Aid Department page.

Q. Can I get scholarships?

Ans. Scholarships are available to our students and come in the form of entrance, undergraduate and graduate awards. Detailed information on scholarships that are offered to Aviation Marketing students is listed on the Entrance Awards and Scholarship page.

Q. What are the costs of books?

Ans. Be prepared to spend approximately $ 400-$600/year on books. Used books are available on campus as well. Be careful to double-check the edition number of the used book you are buying. It may have changed from last year.

Q. What is housing like in Barrie?

Ans. Barrie is a city of 100,000+ people. Housing can be tricky to find if it is left until the last minute. Take a look at the Student Living page for residence and housing information.

Q. Do I have to work in Barrie on co-op?

Ans. Co-op work terms can be done in Barrie, Toronto, out of province or even out of country. If you choose to return to your home province for a co-op job, that is your choice.

AIRLINE WATER SUPPLIES REGARDING SAFE WATER SUPPLY

Q. What is EPA announcing regarding airline water quality?

Ans. EPA is announcing that it has signed agreements (administrative orders upon consent) with 11 major domestic airlines and 13 smaller charter and low-cost airlines to ensure the safety of the drinking water used by their passengers and crew. This announcement is an update to one that was made on 19 January, 2005, when

EPA announced the results of the additional water quality inspections by EPA enforcement officials on 169 randomly selected domestic and international passenger aircraft at 12 airports throughout the U.S.

Q. What are the results of EPA's sampling of aircraft in 2004?

Ans. During the summer of 2004, EPA conducted water quality sampling of one or more galley water taps, water fountains or lavatory faucets on each of 158 aircraft at 7 airports. As EPA announced in September, 2004, 12.7 per cent of those aircraft (20 aircraft) were found to be positive for the presence of total coliform bacteria.

During November-December of 2004, EPA conducted a second round of water quality sampling of 169 aircraft at 12 airports. This second round of monitoring sought water samples from galley water taps as well as lavatory faucets for each aircraft. EPA found that 17.2 per cent of these aircraft (29 aircraft) were total-coliform-positive.

Of the aircraft sampled by EPA in this second round of sampling, 76.3 per cent were US-based. For this round of sampling, the results for foreign-based or US-based aircraft were relatively close. Of the 40 foreign-based carriers sampled during this round, 17.5 per cent (7 aircraft) were total-coliform-positive. By comparison, of the 129 US-based aircraft sampled during this round, 17.1 per cent (22 aircraft) were total-coliform-positive.

Overall, adding together the results of the first and second rounds of EPA sampling of aircraft, EPA has sampled 327 aircraft in 2004. Overall, 15.0 per cent of the aircraft sampled were found to be total-coliform-positive. Overall, 4.3 per cent of the galleys or water fountains sampled were found to be total-coliform-positive; 14.0 per cent of the lavatory faucets sampled were found to be total-coliform-positive. Overall, 49 aircraft were found to be total-coliform-positive. Overall, 76.8 per cent of the aircraft sampled by EPA in this second round of sampling were US-based, and 23.2 per cent of the aircraft. Of the 76 foreign-based carriers sampled overall, 19.7 per cent were total-coliform-positive. By comparison, of the 251 US-based aircraft sampled, 13.5 per cent were total-coliform-positive.

Q. What is EPA doing about this problem?

Ans. EPA has met with airline officials many times to ensure that they are aware of the problem. EPA is negotiating with the industry to increase monitoring of aircraft water systems and ensure that in the event of contamination, contingency plans are immediately implemented. EPA is negotiating agreements with Delta Airlines, JetBlue Airways, Omni Air International and Southwest Airlines. EPA has also begun negotiations with the smaller, regional and charter aircraft carriers to address drinking water quality with orders similar to those reached with the major carriers. The Agency will also negotiate similar orders with foreign flagged carriers to the extent that it has jurisdiction. After two rounds of testing, data showed 15 per cent of the aircraft had water samples that tested positive for total coliform contamination. As a response to these findings, EPA has initiated several actions to ensure that airline water is as safe as any other public drinking water supply. These orders require the carriers to take steps to investigate the problem, take corrective actions, and notify the public. Furthermore, the orders will require each airline to complete testing of their entire fleet within a period of 12 months and to submit a detailed report to EPA on its findings.

In addition, EPA accelerated the rule-making process to develop regulations specifically for water on-board aircraft. At the present time, EPA is working collaboratively with interested stakeholders and other federal agencies that have oversight responsibilities for the industry to develop a rule that will ensure safe drinking water on all aircraft.

Since EPA has not determined the actual sources of contamination, more study and evaluation will be necessary. In 2004, EPA sampled 327 aircraft and notified the airlines of the findings. When sampling identified total coliform in the water of domestic aircraft, that aircraft was disinfected and retested to ensure that the disinfection was effective. In instances where foreign flag aircraft tested positive for total coliform, those airline companies were notified of the positive test result and advised to disinfect and retest the aircraft.

EPA will continue to update its information and advice to the travelling public as soon as new information is available.

Q. Which airlines signed the administrative orders on consent with EPA?

Ans. Administrative orders on consent have been finalized with: AirTran Airways, Alaska Airlines, Aloha Airlines, American Airlines, America West, ATA Airlines, Champion Air, Continental Airlines, Continental Micronesia, Falcon Air Express, Frontier Airlines, Hawaiian Airlines, Miami Air International, Midwest Airlines, North American Airlines, Northwest Airlines, Pace Airlines, Ryan International Airlines, Spirit Airlines, Sun Country Airlines, United Airlines, US Airways, USA 3000 Airlines and World Airways.

Q. What will the orders require the airlines to do?

Ans. The Administrative Orders on Consent require the airlines to implement regular monitoring and disinfection protocols for their entire fleet of aircraft for a period of two years from the effective date of the Order. Specifically, the orders require the airlines to: perform regular monitoring of aircraft water systems; regularly disinfect aircraft water systems and water transfer equipment; undertake corrective action when there is a total coliform positive sample result; provide public notice when there is a total coliform positive sample result; conduct a study of possible sources of contamination that exist outside of the aircraft; and supply information regarding various aspects of its water practices. The results of the study are to be submitted no later than two years from the effective date of the Order.

For example, in the event that an aircraft were to test positive for total coliform at any of the test points on the aircraft, the airline would be required to immediately suspend water service and provide notification, provide bottled water, re-test the aircraft, disinfect the aircraft and test once again to insure that the disinfection was effective. Once the aircraft tests clean, all water service can resume on the plane. The regular establishment of testing protocols will allow EPA to analyze the data and make determinations about possible sources of contamination and how to address them.

Q. What are coliforms?

Ans. Coliforms are a group of closely related bacteria most of which are natural and common inhabitants of the soil and ambient waters (such as lakes and rivers) and in the digestive tracts of humans and other warm-blooded animals.

The presence of total coliform, in and of itself, is not indicative of a health risk. Coliform bacteria will not likely cause illness. However, the presence of coliform bacteria in drinking water indicates that other disease-causing organisms (pathogens) may be present in the water system.

Q. What is *E. coli*?

Ans. *E. coli* is a sub-group of the fecal coliform group. It is found in great quantities in the intestines of people and warm-blooded animals. If total coliform is present in a drinking water sample, EPA requires that it also be tested for *E. coli* or fecal coliform.

Most *E. coli* are harmless. Some strains, however, may cause illness—diarrhea, cramps, nausea, headaches, or other symptoms. The presence of *E. coli* or fecal coliform in a drinking water sample may indicate human or animal fecal contamination—meaning that pathogens may be present.

Q. Is the water on planes unsafe?

Ans. At this time, EPA does not have sufficient data to make broadly applicable, reliable conclusions about water quality on passenger aircraft. What EPA has found in its testing is that coliform contamination is not limited to a specific sub-category of aircraft. It was found on U.S. flag aircraft, foreign flag aircraft, domestic flights, international flights, large aircraft, small aircraft, lavatories and galleys. In order to address this situation, EPA is committed to keeping the American public well-informed of further testing and actions taken, reviewing existing guidance to determine areas where it might be strengthened, concluding agreements with the airlines and taking enforcement actions where warranted.

Q. What should the travelling public do?

Ans. The travelling public may benefit from the information released by EPA when deciding how they use the water that comes from aircraft tanks. Passengers with suppressed immune systems or others concerned should request bottled or canned beverages while on the aircraft and refrain from drinking tea or coffee that does not use bottled water. While boiling water for one minute will remove pathogens from drinking water, the water used to prepare coffee and tea aboard a plane is not generally brought to a sufficiently high temperature to guarantee that pathogens are killed.

Q. Where does the water on passenger airplanes come from?

Ans. In the United States, water loaded aboard aircraft comes from public water systems. The water provided by public water systems is regulated by state and federal authorities. That water may be delivered to the aircraft holding tank via piping from the airport itself or a hose from a water tanker.

Q. What about international flights?

Ans. A significant part of aircraft travel includes international flights. According to the Air Transport Association (ATA), about 90 per cent of ATA member aircraft have the potential to travel internationally. These aircraft may board water from foreign sources which are not subject to EPA drinking water standards.

Q. Who regulates water on passenger airplanes in the United States?

Ans. In the United States, drinking water safety on airlines is jointly regulated by the EPA, Food and Drug Administration (FDA), and Federal Aviation Administration (FAA). EPA regulates the parent systems that supply water to the airports and the drinking water once it is on board the aircraft. FDA has jurisdiction over culinary water (e.g., ice) and the points where aircraft obtain water (e.g., pipes or tankers) at the airport. FAA requires airline companies submit operation and maintenance plans for all parts of the aircraft, including the potable water system.

Q. What is the airlines' role in ensuring safe water on aircraft?

Ans. The regulatory structure for all public water systems, including aircraft, relies upon self-monitoring and reporting of results to the primary agency. The primary agency for aircraft public water systems is EPA.

Q. How is water currently regulated on passenger airplanes?

Ans. The Safe Drinking Water Act (SDWA) regulates water quality in public water systems. Water Supply Guidance 29 was issued in 1986 in an effort to tailor SDWA requirements to address the unique characteristics of Interstate Commerce Carriers, such as aircraft. Under the guidance, ICC operators had been allowed to substitute an operations and maintenance plan for regular monitoring of the vehicle's water system if it was approved by EPA. At this time, EPA is no longer accepting operation and maintenance plans as the ICC programme is being revised.

5

Air Safety Standards, Guidelines and Announcements

AIR SAFETY

Air safety is a broad term encompassing the theory, investigation and categorization of flight failures, and the prevention of such failures through appropriate regulation, as well as through education and training. It can also be applied in the context of campaigns that inform the public as to the safety of air travel. No matter the speed and economy of any mode of transportation, if it is not perceived and demonstrated as safe, it will find few customers and, with few customers, unless it can still be priced to make a profit, the transportation mode will fail and fade from the scene: The dirigibles of the 1920s and 1930s provide a good example of this principle.

Balanced against the speed of travel and the convenience of schedule, transportation by air must overcome various phobias of much of the travelling public: fear of heights, enclosed spaces, surrender of control. Human phobias are not a factor with cargo shipments, but if the shipment doesn't arrive safely, the air carrier will find few customers seeking its service.

Air accidents tend to make national, even international, news. In major airliner accidents, hundreds of passengers may be affected. Add to this the number of family members who will be available at the airports at either end of the flight, ready for interviews, providing pictures of anguish on television news and the task before the industry becomes plain.

Therefore, the entire industry and the government bodies who regulate and support it put a great deal of effort into making air transportation not only appear safe, but demonstrating that it is the safest mode of transportation available.

Institutions

Certification

In most countries, civil aircraft have to be certified by the Civil Aviation Authority (CAA) to be allowed to fly. The major aviation authorities worldwide are the US Federal Aviation Administration (FAA) the European Aviation Safety Agency (EASA) (which provides regulatory advice to the European Union and to a degree supplanted the regulatory bodies of member countries) and the Joint Aviation Authorities (JAA) which advises the CAAs that are members of the European Civil Aviation Conference). FAA, EASA and JAA collaborate on many issues, especially in order to provide streamlined procedure and avoid conflicting or duplicate requirements. FAA and EASA are, in particular, primarily responsible for the certification of the airliners from the two major manufacturers Boeing and Airbus.

Aircraft are certified against standards set out in the code for each CAA. Those codes are very similar and differ primarily in equipment and environmental standards. Regulations on maintenance, repair and operation provide further direction to the owners of the aircraft so that the aircraft continues to meet design standards.

United States

During the 1920s, the first laws were passed in the USA to regulate civil aviation. Of particular significance was the Air Commerce Act, 1926, which required pilots and aircraft to be examined and licensed, for accidents to be properly investigated, and for the establishment of safety rules and navigation aids, under the Aeronautics Branch of the Department of Commerce.

Despite this, in 1926 and 1927 there were a total of 24 fatal commercial airline crashes, a further 16 in 1928, and 51 in 1929 (killing 61 people), which remains the worst year on record at an accident rate of about 1 for every 1,000,000 miles flown. Based on the current numbers flying, this would equate to 7,000 fatal incidents per year.

The fatal incident rate has declined steadily ever since, and, since 1997 the number of fatal air accidents has been no more than 1 for every 2,000,000,000 person-miles flown (e.g., 100 people flying a plane for 1000 miles counts as 100,000 person-miles, making it comparable with methods of transportation with different numbers of passengers, such as one person driving a car for 100,000 miles, which is also 100,000 person-miles), making it one of the safest modes of transport.

Another aspect of safety is protection from attack (discussed below). The terrorist attacks of 2001 are not counted as accidents. However, even if they were counted as accidents they would have added only about 2 deaths per 2,000,000,000 person-miles. Unfortunately, only 2 months later, American Airlines Flight 587 crashed in Queens, NY, killing 256 people, including 5 on the ground, causing 2001 to have a very high fatality rate. Even so, the rate that year including the attacks (estimated here to be about 4 deaths per 1,000,000,000 person-miles), may be relatively safe compared to some other forms of transportation.

Safety improvements have resulted from a wide variety of factors, including improved aircraft design, engineering and maintenance, the evolution of navigation aids, and safety protocols and procedures.

It is often reported that air travel is the safest in terms of deaths per passenger mile. The Transportation Safety Board (2006) reports 1.3 deaths per hundred million vehicle miles for travel by car, and 1.7 deaths per hundred million vehicle miles for travel by air. These are not passenger miles. If an airplane has 100 passengers, then the passenger miles are 100 times higher, making the risk 100 times lower. The number of deaths per passenger mile on commercial airlines between 1995 and 2000 is about 3 deaths per 10 billion passenger miles.

Navigation Aids

One of the first navigation aids to be introduced (in the USA in the late 1920s) was airfield lighting to assist pilots to make landings in poor weather or after dark. The Precision Approach Path Indicator was developed from this in the 1930s, indicating to the pilot the angle of descent to the airfield. This later became adopted internationally through the standards of the International Civil Aviation Organization (ICAO).

With the spread of radio technology, several experimental radio based navigation aids were developed from the late 1920s onwards. These were most successfully used in conjunction with instruments in the cockpit in the form of Instrument Landing Systems (ILS), first used by a scheduled flight to make a landing in a snowstorm at Pittsburgh in 1938. A form of ILS was adopted by the ICAO for international use in 1949.

Following the development of radar in World War II, it was deployed as a landing aid for civil aviation in the form of Ground Control Approach (GCA) systems, joined in 1948 by Distance Measuring Equipment (DME), and in the 1950s by airport surveillance radar as an aid to air traffic control. VHF omnidirectional range (VOR) became the predominate means of route navigation during the 1960s. The ground based VOR stations were often combined with DME at the same site, so that pilots could know both their radials in degrees w.r.t. north to, and their slant range distance to, that beacon.

All of the ground-based navigation aids are rapidly being supplemented by satellite-based aids like GPS, which make it possible for aircrews to know their position with great precision anywhere in the world. With the arrival of Wide Area Augmentation System (WAAS), GPS navigation has become accurate enough for vertical (altitude) as well as horizontal use, and is being used increasingly for instrument approaches as well as en-route navigation. However, since the GPS constellation is a single-point of failure that can be switched off by the U.S. military in time of crisis, ground-based navigation aids are still required for backup.

Air Safety Topics

Lightning

Boeing studies have shown that airliners are struck by lightning on average of twice per year. While the "flash and bang" can be dramatic and startling to the passengers and crew, aircraft are able to withstand normal lightning strikes.

The dangers of more powerful positive lightning were not understood until the destruction of a glider in 1999. It has since been suggested that positive lightning may have caused the crash of Pan Am Flight 214 in 1963. At that time aircraft were not designed to withstand such strikes, since their existence was unknown at the time standards were set.

The effects of normal lightning on traditional metal-covered aircraft are well understood and serious damage from a lightning strike on an airplane is rare. However, as more and more aircraft, like the upcoming Boeing 787, whose whole exterior is made of non-conducting composite materials take to the skies, additional design effort and testing must be made before certification authorities will permit these aircraft in commercial service.

Ice and Snow

Snowy and icy conditions are frequent contributors to airline accidents. The 8 December, 2005 accident where Southwest Airlines Flight 1248 slid off the end of the runway in heavy snow conditions is just one of many examples. Just as on a road, ice and snow buildup can make braking and steering difficult or impossible if severe enough.

The icing of wings is another common problem that is well-known and measures have been developed to combat it. The greatest concern regarding icing is that even a small amount of ice or coarse frost can greatly decrease the ability of a wing to develop lift. This could prevent an otherwise capable aircraft from safely taking off. If ice builds up during flight the result can be catastrophic as evidenced by the crash of American Eagle Flight 4184 (an ATR-72 aircraft) near Roselawn, Indiana on 31 October, 1994, killing 68.

Airlines and airports expend considerable effort to ensure that aircraft are properly de-iced before takeoff whenever the weather threatens to create icing conditions. Modern airliners are designed to prevent ice build-up on wings, engines, and tails by either routing heated air from jet engines through the leading edges of the wing, tail, and inlets, or on slower aircraft, by use of inflatable rubber "boots" that expand and break off any accumulated ice.

Finally, airline dispatch offices keep close watch on weather along the routes of their flights, helping the pilots avoid the worst of possible inflight icing conditions. Pilots can also be equipped with an ice detector in order to leave icy areas they have inadvertently flown into.

Engine Failure

Although aircraft are now designed to fly even after the failure of one or more aircraft engines, the failure of the second engine on one side for example is obviously serious. Losing all engine power is even more serious, as illustrated by the 1970 Dominicana DC-9 air disaster, when fuel contamination caused the failure of both engines. To have an emergency landing place is then very important.

In the 1983 Gimli Glider incident, an Air Canada flight suffered fuel exhaustion during cruise flight, forcing the pilot to glide the plane to an emergency deadstick landing. The automatic deployment of the Ram Air Turbine maintained the necessary hydraulic pressure to the flight controls, so that the pilot was able to land with only a minimal amount of damage to the plane, and minor (evacuation) injuries to a few passengers.

The ultimate form of engine failure, physical separation, occurred in 1979 when a complete engine detached from American Airlines Flight 191, causing damage to the aircraft from which the pilots were unable to recover.

Metal Fatigue

Metal fatigue has occasionally caused failure either of the engine (for example in the 1989 Kegworth Air Disaster), or of the aircraft body, for example the De Havilland Comets in 1953 and 1954 and Aloha Flight 243 in 1988. Now that the subject is better understood, rigorous inspection and non-destructive testing procedures are in place to attempt to identify potential problems.

Delamination

Composite materials consist of layers of fibers embedded in a resin matrix. In some cases, especially when subjected to cyclic stress, the fibers may tear off the matrix, the layers of the material then separate from each other—a process called delamination, and form a mica-like structure which then falls apart. As the failure develops inside the material, nothing is shown on the surface; instrument methods (often ultrasound-based) have to be used.

Numerous modern aircraft have developed delamination problems, but most were discovered before they caused a catastrophic failure. Delamination risk is as old as composite material. Even in the 1940s, several Yak-9s experienced delamination of in their construction.

Stalling

Stalling an aircraft (increasing the angle of attack to a point at which the wings fail to produce enough lift) is a potential danger, but is normally recoverable. Certain devices have been developed to warn the pilot as stall approaches. These include stall warning horns (now standard on virtually all powered aircraft), stick shakers and voice warnings. Two well-known stall-related airline accidents, were the Staines air disaster in 1972, and the United Airlines Flight 553 crash, while on approach to Chicago Midway International Airport, also in 1972.

Fire

Safety regulations control aircraft materials and the requirements for automated fire safety systems. Usually these requirements take the form of required tests. The tests measure flammability and the toxicity of smoke. When the tests fail, they fail on a prototype in an engineering laboratory, rather than in an aircraft.

Fire on board the aircraft, and more especially the toxic smoke generated, have been the cause of several incidents. An electrical fire on Air Canada Flight 797 in 1983 caused the deaths of 23 of the 46 passengers, resulting in the introduction of floor level lighting to assist people to evacuate a smoke-filled aircraft. Two years later a fire on the runway caused the loss of 53 lives, 48 from the effects of smoke, in the 1985 Manchester air disaster. This incident raised serious concerns over the standard aircraft emergency evacuation time of ninety seconds, and calls for the introduction of smoke hoods or misting systems although both were rejected. It did result in the introduction of revised overwing emergency exit doors on certain new aircraft, and a small increase in the spacing between seats next to the emergency exit.

The cargo holds of most airliners are equipped with "fire bottles" (essentially remote-controlled fire extinguishers) to combat a fire that might occur in with the baggage and freight below the passenger cabin. This was due to a terrible accident in 1996. In May of that year a ValuJet DC-9 crashed into the Florida Everglades a few minutes after takeoff after a fire broke out in the forward cargo hold. All 110 aboard were killed.

The investigation determined that improperly packaged chemical oxygen generators (used for the drop-down oxygen masks in the aircraft cabin) had been loaded into the cargo hold. Oxygen generators produce oxygen through a chemical reaction that also generates hundreds of degrees of heat. When installed for use in the ceiling above the passenger seats they are surrounded by heat-resistant shielding and present no fire hazard. On this flight they had been put loosely into a cardboard box for shipment from a maintenance facility. It is likely that one or more of the generators ignited, during or immediately after takeoff, producing an oxygen-rich environment. The cardboard box containing the generators would have quickly caught fire from the heat of the ignited generator. The fire spread to an aircraft tire that was also carried in the hold. Ordinarily the fire would have smothered itself, because of the airtight design of that cargo compartment. But the oxygen generators kept feeding oxygen to the fire, defeating the smothering design of the DC-9 cargo hold. The fire rapidly burned through the passenger cabin floor, incapacitating all aboard with smoke and poisonous gases very quickly. The pilots, although having smoke masks and separate oxygen supplies, had no hope of maintaining control as control cables and electrical wiring burned through.

The maintenance facility (SabreTech) was subjected to large fines and ValuJet, due to this accident and other irregularities, was grounded. The airline reemerged as a smaller airline and eventually merged with Air Tran, a smaller carrier. Adopting the acquired airline's name, the airline has since provided safe service. For the airline industry, rules for the shipment of oxygen generators was severely restricted and cargo holds on larger airliners were required to have "fire bottles" installed.

At one time firefighting foam paths were laid down before an emergency landing, but the practice was considered only marginally effective, and concerns about the depletion of firefighting capability due to pre-foaming led the United States FAA to withdraw its recommendation in 1987.

Bird Strike

Bird strike is an aviation term for a collision between a bird and an aircraft. It is a common threat to aircraft safety and has caused a number of fatal accidents. In 1988 an Ethiopian Airlines Boeing 737 sucked pigeons into both engines during take-off and then crashed in an attempt to return to the Bahir Dar airport; of the 104 people aboard, 35 died and 21 were injured. In another incident in 1995, a Dassault Falcon 20 crashed at a Paris airport during an emergency landing attempt after sucking lapwings into an engine, which caused an engine failure and a fire in the airplane fuselage; all 10 people on board were killed.

Modern jet engines have the capability of surviving an ingestion of a bird. Small fast planes, such as military jet fighters, are at higher risk than big heavy multi-engine ones.

The highest risk of the bird strike is during the takeoff and landing, in low altitudes, which is in the vicinity of the airports. Some airports use active countermeasures, ranging from a person with a shotgun through recorded sounds of predators to employing falconers. Poisonous grass can be planted that is not palatable to birds, nor to insects that attract insectivorous birds.

Passive countermeasures involve sensible land-use management, avoiding conditions attracting flocks of birds to the area (e.g. landfills).

Ground Damage

Aircraft are often victims of damage caused by ground equipment at the airport. In the act of servicing the aircraft between flights a great deal of ground equipment must operate in close proximity to the fuselage and wings. Occasionally the aircraft gets bumped or worse.

Damage may be in the form of simple scratches in the paint or small dents in the skin. However, because aircraft structures (including the outer skin) play such a critical role in the safe operation of a flight, all damage is inspected, measured and possibly tested to ensure that any damage is within safe tolerances. A dent that may look no worse than common "parking lot damage" to an automobile can be serious enough to ground an airplane until a repair can be made.

An example of the seriousness of this problem was the 26 December, 2005 depressurization incident on an Alaska Airlines MD-83 aircraft. During ground services a ramp worker hit the side of the aircraft with a piece of ground equipment. This created a crease in the metal skin. This damage was not reported and the plane departed. Climbing through 26,000 feet the crease in the metal gave way due to the growing difference in pressure between the inside of the aircraft and the outside air. The cabin depressurized with a bang, frightening all aboard and necessitating a rapid descent back to denser (breathable) air and an emergency landing.

The three pieces of ground equipment that most frequently damage aircraft are the passenger boarding bridge, catering trucks, and cargo "beltloaders". However, any other equipment found on an airport ramp can damage an aircraft through careless use, high winds, mechanical failure, and so on.

The generic industry colloquial term for this damage is "ramp rash."

Volcanic Ash

Plumes of volcanic ash near active volcanoes present a risk especially for night flights. The ash is hard and abrasive and can quickly cause significant wear on the propellers and turbocompressor blades, and scratch the cabin windows, impairing visibility. It contaminates fuel and water systems, can jam gears, and can cause a flameout of the engines. Its particles have low melting point, so they melt in the combustion chamber and the ceramic mass then sticks on the turbine blades, fuel nozzles, and the combustors, which can lead to a total engine failure. It can get inside the cabin and contaminate everything there, and can damage the airplane electronics.

There are many instances of damage to jet aircraft from ash encounters. In one of them in 1982, a British Airways Boeing 747 flew through an ash cloud, lost all four engines, and descended from 36,000 feet to only 12,000 feet before the flight crew managed to restart the engines.

With the growing density of air traffic, encounters like this are becoming more common. In 1991 the aviation industry decided to set-up Volcanic Ash Advisory Centers (VAACs), one for each of 9 regions of the world, acting as liaisons between meteorologists, volcanologists, and the aviation industry.

Human Factors

Human factors including pilot error are another potential danger, and currently the most common factor of aviation crashes. Much progress in applying human factors to improving aviation safety was made around the time of World War II by people such as Paul Fitts and Alphonse Chapanis. However, there has been progress in safety throughout the history of aviation, such as the development of the pilot's checklist in 1937. Pilot error and improper communication are often factors in the collision of aircraft. This can take place in the air (1978 PSA Flight 182) or on the ground (1977 Tenerife disaster). The ability of the flight crew to maintain situation awareness is a critical human factor in air safety.

Failure of the pilots to properly monitor the flight instruments resulted in the crash of Eastern Air Lines Flight 401 in 1972, and error during take-off and landing can have catastrophic consequences, for example cause the crash of Prinair Flight 191 on landing, which also in 1972.

Rarely, flight crew members are arrested or subject to disciplinary action for being intoxicated on the job. In 1990, three Northwest Airlines crew members were sentenced to jail for flying from Fargo, North Dakota to Minneapolis-St. Paul International Airport while drunk. In 2001, Northwest fired a pilot who failed a breathalyzer test after flying from San Antonio, Texas to Minneapolis-St. Paul. In July 2002, two America West pilots were arrested just before they were scheduled to fly from Miami, Florida to Phoenix, Arizona because they had been drinking alcohol. The pilots have been fired from America West and the FAA revoked their pilot's licenses. As of 2005 they await trial in a Florida court. The incident created a public relations problem and America West has become the object of many jokes about drunk pilots. While these drunk-flying incidents did not result in crashes, they underscore the role that poor human choices can play in air accidents.

Human factors incidents are not limited to errors by the pilots. The failure to close a cargo door properly on Turkish Airlines Flight 981 in 1974 resulted in the loss of the aircraft—however, the design of the cargo door latch was also a major factor in the incident. In the case of Japan Airlines Flight 123, improper maintenance resulted in the loss of the vertical stabilizer.

Controlled Flight into Terrain (CFIT) is a class of accident in which an undamaged aircraft is flown, under control, into terrain. CFIT accidents typically are a result of pilot error or of navigational system error. Some pilots, convinced that advanced electronic navigation systems such as GPS and INS coupled with Flight Management System computers, or over-relianced on them, are partially responsible for these accidents, have called CFIT accidents "computerized flight into terrain." Failure to protect Instrument Landing System critical areas can also cause controlled flight into terrain. Crew awareness and monitoring of navigational systems can prevent or eliminate CFIT accidents. Crew Resource Management is a modern method now widely used to improve the human factors of air safety. The Aviation Safety Reporting System, or ASRS is another.

Other technical aids can be used to help pilots maintain situational awareness. A ground-collision warning system is an on-board system that will alert a pilot if the aircraft is about to fly into the ground. Also, air traffic controllers constantly monitor flights from the ground and at airports.

Terrorism can also be considered a human factor. Crews are normally trained to handle hijack situations. Prior to the September 11, 2001 attacks, hijackings involved hostage negotiations. After the September 11, 2001 attacks, stricter measures are in place to prevent terrorism using passenger screening technology, air marshals, and precautionary policies. In addition, counter-terrorist organizations monitor potential terrorist activity.

Although most air crews are screened for psychological fitness, some may take suicidal actions. In the case of EgyptAir Flight 990, it appears that the first officer (co-pilot) deliberately dove his aircraft into the Atlantic Ocean while the captain was away from his station, in 1999 off Nantucket, MA, USA. Motivations are unclear, but recorded inputs from the black boxes showed no mechanical problem, no other aircraft in the area, and was corroberated by the voice cockpit recorder.

Certain electronic equipment use is partially or entirely prohibited as it may interfere with aircraft operation, such as causing compass deviations. Use of personal electronic devices and calculators may be prohibited when an aircraft is taking off or landing. The American FCC prohibits use of cell phone on most flights, because inflight use creates problems with the ground-based cells. There is also concern about possible interference with aircraft navigation systems, although that has never been proven to be a serious risk on airliners. A few flights now allow use of cell phones, where the aircraft have been specially wired and certified to meet both FAA and FCC regulations.

Airport Design

Airport design and location can have a big impact on air safety, especially since some airports such as Chicago Midway International Airport were originally built for propeller planes and many airports are in congested areas where it is difficult to meet newer safety standards. For instance, the FAA issued rules in 1999 calling for a Runway Safety Area, usually extending 500 feet to each side and 1000 feet beyond the end of a runway. This is intended to cover ninety per cent of the cases of an aircraft leaving the runway by providing a buffer space free of obstacles. Since this is a recent rule, many airports do not meet it. One method of substituting for the 1000 feet at the end of a runway for airports in congested areas is to install an Engineered Materials Arrestor System, or EMAS. These systems are usually made of a lightweight, crushable concrete that absorbs the energy of the aircraft to bring it to a rapid stop. They have stopped three aircraft (as of 2005) at JFK Airport.

Accidents and Incidents

- Accidents and incidents in aviation.

- Flight data recorder, includes *Cockpit voice recorder*.

Investigators

- Air Accidents Investigation Branch (UK).
- National Transportation Safety Board (USA).
- Transportation Safety Board (Canada).
- Bundesstelle für Flugunfalluntersuchung (Germany).
- Bureau d'Enquêtes et d'Analyses pour la sécurité de l'Aviation Civile (France).
- Australian Transport Safety Bureau (Australia).
- Air Accident Investigation Unit (Ireland) http://www.aaiu.ie/

Regulation

- Federal Aviation Administration
  - Federal Aviation Regulations
- Joint Aviation Authorities
- United Kingdom Civil Aviation Authority
- Irish Aviation Authority—http://www.iaa.ie/
- Transportation Safety Board of Canada.

Aviation.org: A Case Study

Aviation.Org.—the interactive website operated by Aviation Safety Connection, Inc.—addresses fundamental safety issues related to pilot performance. Central themes are embodied in the above-listed key words.

Perspective

The human element is a causal or contributing factor in the great majority of aircraft accidents. Interestingly, the ratio of human factors accidents is relatively constant across the entire aviation spectrum. This percentage—generally accepted at 70 per cent to 80 per cent—does not vary with pilot experience or types of aircraft flown.

Although the magnitude of the problem may be known, the solutions are less easily determined. Nevertheless, each of us can learn by realistically examining our own performances in the cockpit, by learning from our own misjudgments, and by learning from the mistakes of others.

Interactivity

Aviation.Org provides information related to cockpit performance issues and intends to be a catalyst for pilot-to-pilot examination and discussion of these issues. Through this interactive learning process, each of us can strive to attain the high personal and professional standards, the consistent behaviour patterns and the positive attitudes that safe flying requires.

Site Navigation

A summary of Aviation.Org's features and the means to navigate the site are presented in Navigation. Please visit this page to gain a better understanding of this site's content and format.

FAA AVIATION SAFETY DRAFT DOCUMENTS

We invite you to comment on Aviation Safety draft guidance and publication documents. We will provide here draft advisory circulars, orders, notices, policy, technical standard orders, and publications when available for comments. All comments received will be reviewed and considered for incorporation in the final document.

Final Aircraft Certification advisory circulars, other policy documents, and Technical Standard Orders (TSOs) are available on FAA's Regulatory and Guidance Library (RGL).

Document Type	*No. Open*	*Last Updated*
Advisory Circulars	13	02/15/2007
Notices	1	03/01/2007
Orders	none	
Policy	1	02/13/2007
Publications	none	
Technical Standard Orders (TSOs)	5	03/06/2007

AIRCRAFT MANAGEMENT INTERAGENCY COMMITTEE FOR AVIATION POLICY (ICAP) SAFETY STANDARDS AGREEMENT

ICAP Safety Standards Agreement

Fifteen federal agencies that use aircraft have signed a Safety Standards Agreement, developed by GSA and the ICAP. Through this agreement, the agencies consent to adopt the ICAP's Safety Standards Guidelines, write agency-specific standards based on the guidelines, and provide for independent enforcement of the agency standards.

Following is a listing of the agencies that have signed the ICAP Safety Standards Agreement and copy of the document:

Safety Standards Agreement among the Agencies of the ICAP

- Department of Agriculture.
- Department of Commerce.
- Department of Energy.
- Department of Health and Human Services.
- Department of Justice.
- Department of State.
- Department of the Interior.
- Department of the Treasury.
- Department of Transportation/FAA (signed a modified version).
- Department of Veterans Affairs.
- Environmental Protection Agency.
- General Services Administration.
- National Aeronautics and Space Administration.
- National Transportation Safety Board.
- Tennessee Valley Authority.

1. *Background*: The Aircraft Management Policy Advisory Board (AMPAB), chartered by GSA, issued its final report in June, 1998. That report recommends that public aircraft should set the nation's standard for safe and efficient operations. The members of the ICAP and the committee's sponsor, GSA, have determined that this agreement is an important step in meeting the board's recommendation.

Concurrent with the AMPAB's review, the ICAP's Safety Standards and Training Sub-committee developed the "ICAP Safety Standards Guidelines for Federal Flight Programmes" (Safety Standards Guidelines), which the ICAP ratified at their meeting on 16 December, 1998, and which they revised on 22 December, 1999. The guidelines serve as a common framework for each agency to use to develop its own agency-specific flight programme standards for aviation management/administration, operations, maintenance, training and safety.

2. *Purpose*: The agencies listed above enter into this agreement among themselves with the objective of setting the nation's standard for safe and efficient aviation operations.

3. *Basic Terms and Agreements*: Each agency, by signing this agreement, agrees to:

(*a*) Adopt the ICAP Safety Standards Guidelines.
(*b*) Write and publish, as appropriate, agency-specific standards, based on the Safety Standards Guidelines, for aircraft that the agencies own, borrow, bail, or hire (i.e., lease, charter, rent, or otherwise contract for commercially or under inter-service support agreements).
(*c*) Assist each other in implementing the guidelines, and particularly to help those agencies who hire (but do not own) aircraft to tailor the requirements in the guidelines to the agency's special circumstances.
(*d*) Establish a plan and schedule for self-oversight, using independent agents such as, but not limited to—
  (*i*) Inspectors general;
  (*ii*) Contractors;
  (*iii*) Internal review organizations;
  (*iv*) Other bureaus;
  (*v*) Other Federal agencies; or
  (*vi*) The Federal Aviation Administration.

Each agency is solely responsible for managing its own flight programme, for writing its own standards (for both owned and hired aircraft) based on the Safety Standards Guidelines, and for instituting a self-oversight programme that includes independent inspection services, which are obtained and managed by the agency.

(*e*) Continue to facilitate the Aircraft Resources Management Survey (ARMS) programme.
(*f*) Assist each other in obtaining support for inspections and other aviation services.
(*g*) Share best practices.

4. *Duration of Agreement*: This agreement goes into effect for any party on the date that the party signs the agreement and will remain in effect until that party terminates its participation in the agreement.

5. *Termination/Amendment*:

(*a*) Any party may terminate its participation in this agreement, in whole or part, upon written notice to all other parties.

(*b*) Any party may request to modify this agreement by written amendment, and the modification will take effect when the amended agreement is signed by all parties.

6. *Authority*: The parties listed above enter into this agreement pursuant to authority contained in OMB Circular A-126, dated May 22, 1992.

7. Point of Contact

8. Signed.

AIRCRAFT MANAGEMENT SAFETY STANDARDS GUIDELINES FOR FEDERAL FLIGHT PROGRAMMES

ICAP Safety Standards Guidelines

GSA and the members of the Interagency Committee for Aviation Policy (ICAP) have developed and adopted a set of common "Safety Standards Guidelines for Federal Flight Programmes." The ICAP-member agencies have agreed to write and enforce their own agency-specific flight programme standards based on these guidelines.

Note: With the publication of 41 CFR 102-33, "Management of Government Aircraft," on 6 November, 2002, some of the definitions in Section II of these guidelines were superseded. Check the regulation for the correct definitions.

Safety Standards Guidelines for Federal Flight Programmes

The following guidelines were approved by consensus of the ICAP on 16 December, 1998 and were revised on 22 December, 1999.

I. *Preamble*

The guidelines in this document provide federal agencies that own or hire aircraft, except the Department of Defence (DoD), with a framework for developing comprehensive flight programme standards. The ICAP's intent in publishing these guidelines is to enhance operational safety and effectiveness. Each agency should use these guidelines to develop, implement, and maintain agency-specific aviation flight programme standards. Agencies' standards should identify the risks that may not be addressed under the Federal Aviation Regulations (FAR), but which are associated with government flight operations. Where FAR or military standards apply, federal agencies will meet or exceed those standards. Where

FAR or military standards do not apply, federal agencies will use risk management techniques to create the highest standards for themselves (based on the Safety Standards Guidelines described below). Each agency is solely responsible for managing its own flight programme, for writing its own standards (for both owned and hired aircraft) based on the Safety Standards Guidelines, and for instituting a self-oversight programme that includes independent inspection services, obtained and managed by the agency.

II. *Definitions for the Purposes of These Standards*

[*Note*: The definitions in this document should match the definitions in the revised Office of Management and Budget (OMB) Circular A-126 and FPMR 101-37 when those revisions are approved by proper authority. Until those revisions are approved, the following are the ICAP's proposed working definitions.]

(*a*) *Bailed Aircraft*: means aircraft owned by one entity but under the operational control of another entity for an indefinite period under an agreement that may or may not include financial recompense. (In general practice, bailment has come to refer to federal-to-federal transactions. However, an aircraft that a federal agency "borrows" from any entity is considered a bailed aircraft.)

(*b*) *Charter Aircraft*: means an aircraft procured under a one time exclusive agreement that specifies performance and payment. The vendor operates charter aircraft.

(*c*) *Contract Aircraft*: means an aircraft procured for an agency's exclusive use for a specified period of time by means of a formal contract under which the contractor is responsible for the operation, safety, and maintenance of the aircraft.

(*d*) *Owned Aircraft*: means an aircraft for which title or rights of title are vested with a federal entity.

(*e*) *Fleet Aircraft*: Refers to any of the following:

1. Owned aircraft.
2. Aircraft under lease/purchase agreement where the intent is to transfer title to the federal government.
3. An aircraft leased in lieu of ownership for an agreed period of 180 or more days.
4. Any aircraft under exclusive use and operational control of a federal entity for 180 or more days-

excludes aircraft operating under full service contracts.

(*f*) *Leased Aircraft*: means an aircraft that the government has exclusive right (through a financial contract) to use for a specific period of time and that the procuring agency is responsible for the operation and safety of the aircraft. Maintenance responsibility is defined under terms of the contract.

(*g*) *Loaned Aircraft*: means an aircraft owned by one entity, but under the operational control of another entity for a limited period of time under an agreement that does not include financial recompense. (In general practice, a loaned aircraft is owned by a federal agency and is operated by a state, municipality, or other non-federal government entity on a temporary basis. To loan means to furnish property to another for temporary use with provisions that the property will be returned in essentially the same condition and without payment.)

III. ***Applicability***

These guidelines do not supersede or relieve any agency of the requirement to comply with any existing U.S. and U.S.-ratified international laws, regulations, policies and directives.

This document and the safety standards guidelines set forth in this paragraph IV apply to the following:

(*a*) Federal agencies who operate owned, leased, bailed, or fleet aircraft (as defined within this document), exclusively in the service to the federal government for the conduct of governmental missions, should establish comprehensive flight programme standards based on the guidelines described in paragraph IV below.

(*b*) Agencies who charter or contract for aircraft (as defined by this document), exclusively in the service to the federal government for the performance of a governmental mission, should require the vendor or contractor to comply with (1) the civil standards applicable to the type of operation conducted and/or, (2) comprehensive flight programme standards based on the guidelines described in this document.

This document and the safety standards guidelines set forth in paragraph IV of this document do not apply to the following:

(*a*) Aircraft owned by the armed forces or operated on behalf of the United States Government by armed forces personnel (including the Reserves) and to aircraft owned by the National Guard or operated by National Guard personnel for the National Guard of a State, Territory, Puerto Rico, or the District of Columbia, or Civil Reserve Air Fleet Programme.

(*b*) Loaned aircraft (i.e., owned by a federal agency but operated by and on behalf of the loanee) unless (1) the aircraft is used for a federal government mission, and (2) requirements for applying these standards are established in a Memorandum of Agreement governing the loan.

IV. *Comprehensive Safety Standards Guidelines to Promote Safety and Reduce or Eliminate Accidents*

Federal agencies should apply the following standards guidelines to all areas of their flight programmes, including management/administration, operations, maintenance, training and safety. In each of those areas, agencies should implement agency-specific standards by developing written policies and procedures (as appropriate for the sizes and types of their aircraft and scopes of operations of their missions) for the following elements:

A. Management/Administration

1. A management structure responsible for the administration, operation, safety, training, maintenance, and financial needs of the operation;
2. Roles, responsibilities, and authorities of assigned managers, pilots, maintenance personnel, flight crewmembers, flight safety personnel, and dispatchers (as applicable);
3. Record keeping procedures to record and track flight crewmember flight and duty time and training;
4. Record keeping procedures to record and track maintenance personnel duty time and training.

B. Operations

1. Basic qualifications and currency requirements for the pilots, crewmembers, maintenance personnel, and other

mission related personnel, as required by (applicable to) the agency's mission;

2. Pilot and crewmember duty time and flight time limitations;
3. Compliance with owning agency or military safety of flight notices and operational bulletins;
4. Flight following to provide for timely notification of management and initiating search and rescue operations in case of a lost or downed aircraft;
5. Appropriate emergency procedures and equipment and, if Aviation Life Support Equipment (ALSE) is required by the agency for a specific mission, then a programme for ensuring the ALSE is inspected and serviceable.

C. Maintenance

1. Aircraft maintenance and inspection programmes to ensure safety of flight in accordance with either applicable military programmes, manufacturers' programmes, agency-prescribed programmes, or civil programmes;
2. Compliance with owning agency or military safety of flight notices, Federal Aviation Authority (FAA) airworthiness directives, or mandatory manufacturers' bulletins applicable to the types of aircraft, engine(s), propeller(s) and appliances;
3. Operation of aircraft with inoperable equipment;
4. Applicable technical support, including appropriate engineering documentation and testing, for aircraft, power plant, propeller, or appliance repairs, modifications, or equipment installations;
5. Quality control for the purchase and acquisition of replacement parts, assuring that parts purchased or acquired have the necessary documentation to determine airworthiness;
6. Record keeping procedures to record and track maintenance actions; inspections; retirement life components and parts; and the flight hours, cycles, and calendar times for Flight Safety Critical Aircraft Parts (FSCAP, i.e., DoD surplus/excess).

D. Training

1. Initial and recurrent training appropriate for the responsibilities and necessary operational skills of all aviation personnel and relevant to the types of operations/ missions conducted by the agency.

E. Safety

1. Risk analysis and risk management to identify and mitigate hazards and provide procedures for managing risk to an acceptable level; internal and external assessments and oversight to verify that the standards elements called for by this document are implemented;
2. A system for communicating and reporting hazards, incidents, and accidents and disseminating safety/accident prevention (and other) information;
3. An accident response plan;
4. Notification to National Transportation Safety Board (NTSB) and GSA of accidents and incidents defined by Title 49 CFR Part 830;
5. Aviation safety awards programme.

NATIONAL TRANSPORTATION SAFETY BOARD

The National Transportation Safety Board (NTSB) is a U.S. Government independent organization responsible for investigation of accidents involving aviation, highway, marine, pipelines and railroads in the United States. It is charged by the U.S. Congress to investigate every civil aviation accident in the United States, as well as significant accidents in other modes of transportation (such as the Big Bayou Canot train disaster near Mobile, Alabama). The organization is also in-charge of investigating cases of hazardous waste releases that occur from modes of transportation. Mark Rosenker was appointed as Vice Chairman in 2003 and Acting Chairman in March, 2005. He has held the position of the Chairman since August, 2006. Robert Sumwalt is the vice-chairman. The board is based in Washington, D.C.

History

It was established as an independent organization in 1967, taking over the investigative functions of the Civil Aeronautics Board, among other duties. Originally established with strong ties to the

US Department of Transportation, these ties were later severed under the Independent Safety Board Act of 1975. The organization receives its authority from Chapter 11, Title 49 of the United States Code. It has investigated over 124,000 aviation incidents since its establishment.

Organization

The board has five members, one of whom is designated the chairman by the U.S. President and then approved by the U.S. Senate for a fixed 2-year term. Another member is designated as vice-chairman, and becomes acting chairman when there is no formal chairman. Organization within the Board is composed of separate sub-offices for highway safety, maritime safety, aviation safety, railroad, pipeline, and hazardous material investigations, research and engineering, recommendations and communications, academy and administrative law judges. These sub-offices report to the Office of Management.

Investigations

The NTSB is normally the lead organization in the investigation of a transportation accident within its sphere. However, this power can be surrendered to other organizations if the US Attorney General declares the case to be linked to an intentional criminal act. This occurred during the investigation of the 11 September, 2001 -attacks, when the US Justice Department took over the investigation.

An investigation of an incident within the United States typically starts with the creation of a "go team", composed of specialists in fields relating to the incident. This is followed by the designation of other organizations or corporations as parties to the investigation. The Board may then choose to hold public hearings on the issue. Finally, it will compose a final statement and may issue safety recommendations.

The NTSB may investigate incidents or accidents occurring outside the United States under certain circumstances. These may include:

- accidents or incidents occurring to American-registered or American-owned aircraft in foreign airspace if the aircraft both departed and was scheduled to land in the United States. This has happened on rare occasion with respect to flights to and within Alaska that have crashed in Canada.

- accidents or incidents occurring to American-registered or American-owned aircraft in countries without a transportation investigative board.

The NTSB, if asked, will also provide technical and other advice for a fee to transportation investigative boards in countries that do not have the equipment or specialized technicians available to undertake all aspects of a complex investigation.

Accidents and Incidents

The NTSB has investigated thousands of accidents and incidents since it was formed in 1967. These have included: (See Table 5.1 on page 201).

AIRLINE WATER SUPPLIES

EPA is responsible for safe drinking water, both from your tap and on airplanes. In the summer of 2004, and again that fall, we tested drinking water aboard hundreds of randomly selected domestic and international passenger aircraft. The summer data showed that 13 per cent of tested aircraft water failed to meet our standards. Fall testing showed that 17 per cent failed the standards.

Passengers with compromised immune systems or others concerned may want to request canned or bottled beverages and avoid drinking coffee, tea, and other drinks prepared with tap water. We will update this information and our advice to the traveling public as new information becomes available.

On 9 November, 2004, we announced commitments from 12 major U.S. passenger airlines to implement new aircraft water testing and disinfection protocols. These protocols will further protect the travelling public while we review existing guidance governing potable water aboard passenger aircraft.

Regulating Water on Interstate Carrier Conveyances

In the United States, drinking water safety on airlines is jointly regulated by the EPA, Food and Drug Administration (FDA), and Federal Aviation Administration (FAA).

EPA regulates the parent systems that supply water to the airports and the drinking water once it is on board the aircraft. EPA regulates water quality in public water systems in accordance with the Safe Drinking Water Act (SDWA). The regulatory structure for all public water systems, including aircraft, relies upon self-monitoring and reporting of results to the primacy agency. The primacy agency for aircraft public water systems is EPA.

EPA issued Water Supply Guidance 29 (24 K PDF File, 12 pgs) (ALL ABOUT PDF FILES) in 1986 in an effort to tailor Safe Drinking Water Act requirements for public water systems to address the unique characteristics of Interstate Carrier Conveyances (ICCs), such as aircraft. Under the guidance, ICC operators could substitute an EPA approved operations and maintenance plan for regular monitoring of the vehicle's water system. EPA has initiated an accelerated rule-making process to develop regulations for water on-board aircraft. At this time, EPA is no longer approving operation and maintenance programmes in lieu of monitoring while the ICC programme is being revised.

FDA has jurisdiction over culinary water (e.g., ice) and the points where aircraft obtain water (e.g., pipes or tankers) at the airport:

- FDA regulations for Interstate Conveyance Sanitation.
- FDA Guide to Inspections of Interstate Carriers and Support Facilities.

FAA requires airline companies to submit operation and maintenance plans for all parts of the aircraft, including the potable water system.

Development of Aircraft Drinking Water Regulations

EPA has accelerated the development of regulations for water that is served onboard aircraft. The new regulations will tailor drinking water rule requirements to the unique characteristics of aircraft.

EPA's objective is to ensure that water passengers drink while on an airplane is as safe as the drinking water they receive from their public water system. In developing the regulations, EPA will utilize a collaborative process with representatives from industry, government and the interested public.

As part of this collaborative process, EPA is holding public meetings and workshops. The times, locations, and past meetings and additional public workshops to identify issues and regulatory options for the aircraft drinking water rule will be made available on the public meetings page.

Table 5.1

Date	Location	Operator	Make/Model	Fatal	Surviving
01/13/82	Washington, D.C.	Air Florida	Boeing 737-222	70	4
01/23/82	Boston, MA	World Airways	McDonnell Douglas DC-10-30	2	198
07/09/82	New Orleans, LA	Pan Am	Boeing 727-235	137	0
11/08/82	Honolulu, HI	Pan Am	Boeing 747-100	1	274
01/09/83	Brainerd, MN	Republic Airlines	Convair 580-11-A	1	29
10/11/83	Pinckneyville, IL	Air Illinois	Hawker Siddeley HS-748-2A	7	0
01/01/85	La Paz, Bolivia	Eastern Air Lines	Boeing 727-225	21	0
01/21/85	Reno, NV	Galaxy Airlines	Lockheed 188C	64	1
08/02/85	Dallas/FT Worth, TX	Delta Air Lines	Lockheed L-1011-385-1	126	26
09/06/85	Milwaukee, WI	Midwest Express Airlines	Douglas DC-9-14	27	0
12/12/85	Gander, Newfoundland	Arrow Airways	Douglas DC-8-63	248	0
02/04/86	Near Athens, Greece	Trans World Airlines	Boeing 727-231	4	110
02/14/87	Durango, MX	Ports of Call	Boeing 707-323B	1	125
08/16/87	Romulus, MI	Northwest Airlines	McDonnell Douglas DC-9-82	148	1
11/15/87	Denver, CO	Continental Airlines	McDonnell Douglas DC-9-14	25	52
12/07/87	San Luis Obispo, CA	Pacific Southwest Airlines	British Aerospace BAe-146-200	38	0
08/31/88	Dallas/FT Worth, TX	Delta Air Lines	Boeing 727-232	12	89
12/21/88	Lockerbie, Scotland	Pan Am	Boeing 747-121	243	0
02/08/89	Santamaria, Azores	Independent Air	Boeing 707	137	0
02/24/89	Honolulu, HI	United Airlines	Boeing 747-122	9	328
04/28/89	Maui, HI	Aloha Airlines	Boeing 737-297	1	94
07/19/89	Sioux City, IA	United Airlines	McDonnell Douglas DC-10-10	110	175
09/20/89	Flushing, NY	USAir	Boeing 737-400	2	55
12/27/89	Miami, FL	Eastern Air Lines	Boeing 727-225B	1	46
10/03/90	Cape Canaveral, FL	Eastern Air Lines	McDonnell Douglas DC-9-31	1	90
12/03/90	Romulus, MI	Northwest Airlines	McDonnell Douglas DC-9-14	7	33

(Contd.)

Date	Location	Operator	Make/Model	Fatal	Surviving
02/01/91	Los Angeles. CA	USAir	Boeing 737-300	20	63
03/03/91	Colorado Spgs, CO	United Airlines	Boeing 737-291	20	0
03/22/92	Flushing, NY	USAir	Fokker 28-4000	25	22
07/02/94	Charlotte, NC	USAir	Douglas DC-9-30	37	20
09/08/94	Aliquippa, PA	USAir	Boeing B-737-300	127	0
10/31/94	Roselawn, IN	American Eagle	Atr-72-212	64	0
12/20/95	Cali, Colombia	American Airlines	Boeing B-757	152	4
05/11/96	Miami, FL	ValuJet Airlines	McDonnell Douglas DC-9	110	0
07/06/96	Pensacola, FL	Delta Air Lines	McDonnell Douglas MD-88	2	140
07/17/96	Moriches, NY	Trans World Airlines	Boeing 747	230	0
08/02/97	Lima, Peru	Continental Airlines	Boeing 757-200	1	141
12/28/97	Pacific Ocean	United Airlines	Boeing 747	1	373
06/01/99	Little Rock, AR	American Airlines	McDonnell Douglas MD-80	10	129
01/31/00	Point Mugu, CA	Alaska Airlines	McDonnell Douglas MD-83	83	0
09/11/01	New York City, NY	American Airlines	Boeing 767-200	81	0
09/11/01	New York City, NY	United Airlines	Boeing 767-200	56	0
09/11/01	Arlington, VA	American Airlines	Boeing 757-200	58	0
09/11/01	Shanksville, PA	United Airlines	Boeing 757	37	0
11/12/01	Belle Harbor, NY	American Airlines	Airbus Industries A 300-600	251	0
01/08/03	Charlotte, NC	US Airways Express	Beech 1900	19	0
10/19/04	Kirksville, MO	Corporate Airlines	British Aerospace Jetstream 32	13	2
08/27/06	Lexington, KY	Comair	Bombardier Canadair CRJ-100ER	49	1
10/11/06	New York, NY	Cirrus SR 20 (Private)	Cirrus SR 20 N9 29CD	2	0

Basic Information on Airline Water Supplies

EPA is responsible for ensuring that public water systems provide water that is safe to drink, including water from aircraft in kitchens, drinking water fountains and bathroom sinks.

As part of enforcement activities, EPA conducted two rounds of random tests of the water supplies on more than 300 domestic and international aircraft arriving at selected U.S. airports. The first and second rounds of testing showed total coliform in 13 and 17 per cent of samples (respectively). Total coliform are indicators that other disease-causing organisms (pathogens) could be in the water and could potentially affect people's health.

Since releasing the first test results, EPA worked with the airlines to:

- develop agreements to immediately reduce health risks to passengers; and
- provide additional testing to help the Agency determine the nature and extent of the problem.

On 19 October, 2005, EPA announced that agreements (administrative orders on consent) had been finalized with 24 U.S. passenger airlines to implement new aircraft water testing and disinfection protocols. These protocols will protect the travelling public while we review existing regulations governing potable water aboard passenger aircraft. EPA is negotiating with three other major airlines, and has begun working with smaller, regional and charter airlines.

EPA is making information available to help the travelling public make informed decisions. Passengers, with compromised immune systems or others concerned may want to request canned or bottled beverages. EPA will post-updates as new information is available at www/epa.gov/airlinewater.

Under the agreements, airlines will implement quarterly disinfection of water delivery systems aboard passenger aircraft. They will also increase monitoring and notify the public when testing reveals water that does not meet EPA standards. Airlines will also be required to analyze possible sources of contamination that exist outside of the aircraft and to provide information related to practices of boarding water from foreign public water supplies, which are not regulated by EPA.

EPA began a review of existing guidance in 2002. In response to the aircraft test results, EPA is accelerating efforts to develop regulations for aircraft drinking water. The proposed rule is scheduled to be published in 2007, with the final rule issued in 2009. We will work with other federal agencies overseeing the airline industry, industry representatives, and the public. Agency agreements with the airlines and resulting administrative orders signed by the airlines will govern airline drinking water safety for the interim.

Airline Drinking Water Enforcement

EPA is responsible for ensuring the safety of drinking water in the United States under the Safe Drinking Water Act. This includes drinking water on aircraft. In 2004, EPA randomly tested the drinking water on 327 U.S. and foreign flag aircraft from different airlines at 19 airports around the country. The results of that testing showed that 15 per cent of the aircraft had water that tested positive for total-coliform-bacteria. Total coliform is an indicator that other disease-causing organisms (pathogens) could be in the water and could potentially affect people's health.

EPA immediately began working with domestic airlines to address this problem. On 19 October, 2005, EPA announced that settlements had been reached with 24 domestic airlines to implement new aircraft water testing and disinfection protocols. EPA will continue to work with smaller, regional and charter airlines to reach similar agreements to ensure drinking water quality.

Each of the airlines will implement regular monitoring and disinfection protocols for its entire fleet of aircraft for a period of two years. The airlines will:

- regularly monitor aircraft water systems;
- regularly disinfect aircraft water systems and water transfer equipment;
- undertake corrective action when testing reveals a total coliform sample;
- provide public notice or suspend its water service when there is a total coliform positive sample result;
- conduct a study of possible sources of contamination that exist outside of the aircraft; and
- supply information regarding various aspects of its domestic and foreign water practices.

For more information, see a Summary of Administrative

Orders on Consent or visit the Airline Water Supplies page, which provides a summary of the development of a rule for airline drinking water and the 2004 testing results.

JFIM: TERRORISM RELATED INFORMATION

As international aviation makes the world a smaller place, it also puts more people at risk to the unreasonable actions of terrorists. This page has links to take the user to various anti-terrorism websites which help to educate the international traveller, thus enhancing personal safety.

Case Study: Public Announcement by U.S. Department of State, Office of the Spokesman. *This information is current as of today.*

Worldwide Caution

11 October, 2006

This Public Announcement updates information on the continuing threat of terrorist actions and violence against Americans and interests overseas. This supersedes the Worldwide Caution dated 6 June, 2006 and expires on 9 April, 2007.

The Department of State remains concerned about the continued threat of terrorist attacks, demonstrations and other violent actions against U.S. citizens and interests overseas. Current information suggests that al-Qaida and affiliated organizations continue to plan terrorist attacks against U.S. interests in multiple regions, including Europe, Asia, Africa and the Middle East. These attacks may employ a wide variety of tactics to include assassinations, kidnappings, hijackings and bombings.

On-going events in Iraq and elsewhere in the Middle East have resulted in demonstrations and associated violence in several countries. Americans are reminded that demonstrations and rioting can occur with little or no warning.

In August, 2006, British authorities arrested a significant number of extremists engaged in a plot to destroy multiple passenger aircraft flying from the United Kingdom to the United States. The September, 2006 attack on the U.S. Embassy in Syria and the March, 2006 bombing near the U.S. Consulate in Karachi, Pakistan, illustrate the continuing desire of extremists to strike American targets.

Extremists may elect to use conventional or non-conventional weapons, and target both official and private interests. A series of bombings in Thailand in May and September, 2006 which targeted tourist destinations, the bombings in the Egyptian resort town of Dahab in April, 2006, the bombings at a market in Palu, Indonesia in December, 2005, the hotel bombings in Jordan in early November 2005 and the bombing of a crowded marketplace in New Delhi in October, 2005 illustrate how terrorists exploit vulnerabilities associated with soft targets. Additional examples of such targets include high-profile sporting events, residential areas, business offices, clubs, restaurants, places of worship, schools, public areas and locales where Americans gather in large numbers, including during holidays. Financial or economic targets of value may also be considered as possible venues; the failed attack on the Abqaiq oil processing facility in Saudi Arabia in late February, 2006 is one such example.

In the wake of the August, 2006 plot against aircraft in London, numerous terrorist attacks on trains in India in 2006, the July, 2005 London Underground bombings, and the March, 2004 train attacks in Madrid, Americans are reminded of the potential for terrorists to attack public transportation systems. In addition, extremists may also select aviation and maritime services as possible targets.

U.S. citizens are strongly encouraged to maintain a high level of vigilance, be aware of local events, and take the appropriate steps to bolster their personal security. For additional information, please refer to "A Safe Trip Abroad" found at http://travel.state.gov/.

U.S. Government facilities worldwide remain at a heightened state of alert. These facilities may temporarily close or periodically suspend public services to assess their security posture. In those instances, U.S. embassies and consulates will make every effort to provide emergency services to U.S. citizens. Americans abroad are urged to monitor the local news and maintain contact with the nearest U.S. embassy or consulate.

As the Department continues to develop information on any potential security threats to U.S. citizens overseas, it shares credible threat information through its Consular Information Programme documents, available on the Internet at http://travel.state.gov/. In addition to information on the Internet, travellers may obtain up-to-date information on security conditions by calling 1-888-407-4747 toll-free in the U.S. or outside the U.S. and Canada on a regular toll line at 1-202-501-4444.

Other Government Websites

- USA Gov Service Locator.

- Department of Homeland Security.
- The White House.
- more >.

This site is managed by the Bureau of Consular Affairs, U.S. Department of State. External links to other Internet sites should not be construed as an endorsement of the views contained therein. Copyright Information Disclaimers

A Chronology of Some of the World's Recent Air Disasters

2006

29 September: A Boeing 737 carrying 154 passengers and crew crashed into the Amazon rainforest in Brazil, killing all on board, after colliding with a private jet in mid-air.

22 August: A Russian Tupolev-154 passenger plane with 170 people on board crashes north of Donetsk, in eastern Ukraine.

9 July: A Russian S7 Airbus A-310 skids off the runway during landing at Irkutsk airport in Siberia. At least 124 people on board die, but more than 50 survive the crash.

3 May: An Armavia Airbus A-320 crashes into the Black Sea near Sochi, killing all 113 people on board.

2005

10 December: A Sosoliso Airlines DC-9 crashes in the southern Nigerian city of Port Harcourt, killing 103 people on board.

6 December: A C-130 military transport plane crashes on the outskirts of the Iranian capital Tehran, killing 110 people, including some on the ground.

22 October: A Bellview airlines Boeing 737 carrying 117 people on board crashes soon after take-off from the Nigerian city of Lagos, killing everyone on board.

5 September: A Mandala Airlines plane with 112 passengers and five crew on board crashes after take-off in the Indonesian city of Medan, killing almost all on board and dozens on the ground.

23 August: A Tans airline Boeing 737-200 crashes on an internal flight in Peru, near the city of Pucallpa, with at least 40 people reported dead.

16 August: A Colombian plane operated by West Caribbean Airways crashes in a remote region of Venezuela, killing all 160 people on board. The airliner, heading from Panama to Martinique, was packed with residents of the Caribbean island.

14 August: A Helios Airways flight from Cyprus to Prague with 121 people on board crashes north of the Greek capital Athens, apparently after a drop in cabin pressure.

16 July: An Equator plane crashes soon after take-off from Equatorial Guinea's island capital, Malabo, west of the mainland, killing all 60 people on board.

3 February: The wreckage of Kam Air Boeing 737 flight is located in high mountains near the Afghan capital Kabul, two days after the plane vanished from radar screens in heavy snowstorms. All 104 people on board are feared dead.

2004

21 November: A passenger plane crashes into a frozen lake near the city of Baotou in the Inner Mongolia region of northern China, killing all 53 on board and two on the ground, officials say.

24 August: Chechen suicide bombers destroy two Russian airliners carrying a total of 90 passengers and crew on domestic flights out of Moscow.

3 January: An Egyptian charter plane belonging to Flash Airlines crashes into the Red Sea, killing all 141 people on board. Most of the passengers are thought to be French tourists.

2003

25 December: A Boeing 727 crashes soon after take-off from the West African state of Benin, killing at least 135 people en-route to Lebanon.

8 July: A Boeing 737 crashes in Sudan shortly after take-off, killing 115 people on board. Only one passenger, a small child, survived.

26 May: A Ukrainian Yak-42 crashes near the Black Sea resort of Trabzon in north-west Turkey, killing all 74 people on board—most of them Spanish peacekeepers returning home from Afghanistan.

8 May: More than 120 people are feared dead in DR Congo after the rear ramp of an old Soviet plane, an Illusion 76 cargo plane, apparently falls off, sucking them out.

6 March: An Algerian Boeing 737 crashes after taking off from the remote Tamanrasset airport, leaving up to 102 people dead.

19 February: An Iranian military transport aircraft carrying 276 people crashes in the south of the country, killing all on board.

8 January: A Turkish Airlines plane with 76 passengers and crew on board crashes while coming in to land at Diyarbakir.

2002

23 December: An Antonov 140 commuter plane carrying aerospace experts crashes in central Iran, killing all 46 people aboard.

The delegation had been due to review an Iranian version of the same plane built under licence.

27 July: A fighter jet crashes into a crowd of spectators in the west Ukrainian town of Lviv, killing 77 people, in what is the world's worst air show disaster.

1 July: Seventy-one people, many of them children, die when a Russian Tupolev 154 aircraft on a school trip to Spain collides with a Boeing 757 transport plane over southern Germany.

25 May: A Boeing 747 belonging to Taiwan's national carrier—China Airlines—crashes into the sea near the Taiwanese island of Penghu, with 225 passengers and crew on board.

7 May: China Northern Airlines plane carrying 112 people crashes into the sea near Dalian in north-east China.

7 May: On the same day, an EgyptAir Boeing 735 crash lands near Tunis with 55 passengers and up to 10 crew on board. Most people survive.

4 May: A BAC 1-11-500 plane operated by EAS Airlines crashes in the Nigerian city of Kano, killing 148 people-half of them on the ground.

15 April: Air China flight 129 crashes on its approach to Pusan, South Korea, with over 160 passengers and crew on board.

12 February: A Tupolev 154 operated by Iran Air crashes in mountains in the west of Iran, killing all 117 on board.

29 January: A Boeing 727 from the Ecuadorean TAME airline crashes in mountains in Colombia, killing 92 people.

2001

12 November: An American Airlines A-300 bound for the Dominican Republic crashes after takeoff in a residential area of the borough of Queens, New York, killing all 260 people on board and at least five people on the ground.

8 October: A Scandinavian Airlines System (SAS) airliner collides with a small plane in heavy fog on the runway at Milan's Linate airport, killing 118 people.

4 October: A Russian Sibir Airlines Tupolev 154 enroute from Tel Aviv to Novosibirsk in Siberia explodes in mid-air and crashes into the Black Sea, killing 78 passengers and crew.

11 September: Two passenger planes hijacked by Islamic militants are rammed into the World Trade Center in New York. A third is flown into the Pentagon and a fourth crashes south east of Pittsburgh. In total more than 3,000 people killed.

3 July: A Russian Tupolev 154, en-route from Yekaterinburg in the Ural mountains to the Russian port of Vladivostok, crashes near the Siberian city of Irkutsk, killing 133 passengers and 10 crew.

2000

30 October: A Singapore Airlines Boeing 747 bound for Los Angeles crashes after take-off from Taipei airport in Taiwan, killing 78 of the 179 people on board.

23 August: A Gulf Air Airbus crashes into the sea as it comes in to land in Bahrain, killing all 143 people on board.

25 July: Air France Concorde en-route for New York crashes into a hotel outside Paris shortly after takeoff, killing 113 people, including four on the ground.

17 July: Alliance Air Boeing 737-200 crashes into houses attempting to land at Patna, India, killing 51 people on board and four on the ground.

19 April: Air Philippines Boeing 737-200 from Manila to Davao crashes on approach to landing, killing all 131 people on board.

31 January: Alaska Airlines MD-83 from Mexico to San Francisco plunges into ocean off southern California, killing all 88 people on board.

30 January: Kenya Airways A-310 crashes into Atlantic Ocean shortly after takeoff from Abidjan, Ivory Coast, en-route for Lagos, Nigeria. All but 10 of the 179 people on board die.

1999

31 October: EgyptAir Boeing 767 crashes into Atlantic Ocean after taking off from John F. Kennedy Airport in New York on flight to Cairo, Egypt, killing all 217 on board.

24 February: China Southwest Airlines plane crashes in a field in China's coastal Zhejiang province after a mid-air explosion. All 61 people on board the Russian-built TU-154 flying from Chongqing to the south-eastern city of Wenzhou are killed.

1998

11 December: Thai Airways International A-310 crashes on a domestic flight during its third attempt to land at Surat Thani, Thailand, killing 101 people.

2 September: Swissair MD-11 from New York to Geneva crashes in the Atlantic Ocean off Canada killing all 229 people on board.

16 February: Airbus A-300 owned by Taiwan's China Airlines crashes near Taipei's Chiang Kai-shek airport while trying to land in fog and rain after a flight from Bali, Indonesia. All 196 on board and seven people on ground are killed.

2 February: Cebu Pacific Air DC-9 crashes into mountain in southern Philippines, killing all 104 people aboard.

Chronology of Nigerian Air Disasters

20 November, 1969: Nigeria Airways BAC VC 10 crashes on landing killing 87 on board.

22 January, 1973: Royal Jordanian Airlines flight 707 carrying 171 Nigerian Muslims returning from Mecca and 5 crewmen died in crash in Kano, Nigeria.

1 March, 1978: Nigeria Airways F 28-1000 crashes in Kano killing 16.

28 November, 1983: Nigeria Airways F 28-1000 crashes on approach in Enugu killing 53 on board.

11 July, 1991: Nigeria Airways DC-8-61 crashes in Jeddah, Saudi Arabia from system failure killing 261 on board.

26 September, 1992: Nigerian Air Force A C-130 plane crashes minutes after take-off from Lagos. All 200 on board killed.

24 June, 1995: Harka Air Services Tupolev 34 crashes on landing in Lagos killing 16.

13 November, 1995: Nigeria Airways Boeing 737-2F 9 crashes on landing in Kaduna killing 9.

7 November, 1996: A Nigerian ADC (Aviation Development Corporation) Airline Boeing 727-231 flying from Port Harcourt to Lagos with 142 passengers and 9 crew members crashed on landing, plunging into a lagoon with all on board killed.

31 January, 1997: SkyPower Express Airways Embraer 110PIA crashes on landing in Yola killing 5.

5 January, 2000: SkyPower Express Airways Bandeirante 110P 1A crashes on landing in Abuja killing 17.

May 4, 2002: Nigerian EAS Airlines' BAC 1-11-500 with 105 people on board crashed and burst into flames in a poor, densely populated suburb of Kano. 76 on board killed, including 72 on the ground bringing it to a total 148 dead.

22 ***October, 2005:*** A Nigerian Bellview Airlines Boeing 737 airliner with 117 people on board crashes and disintegrates in flames shortly after take-off from Lagos. All on board killed.

10 December, 2005: A Nigerian Sosoliso Airlines DC-9 crashes in Port Harcourt, killing all 103 on board. Most on board were school children going home for Christmas.

17 September, 2006: A Nigerian 18-seater Dornier 228 Air Force transport plane, carrying 15 senior army officers and three crew members crashed leaving only three survivors that sustained serious injuries. The plane with registration number 228-212 crashed in northern part of the country at Oko village in Vandeikya local government area of Benue State at about 10:30 a.m.

29 ***October, 2006:*** A Nigerian ADC (Aviation Development Corporation) Airline Boeing 737 with 104 on board crashes minutes after take-off from Abuja's airport during a rain storm. All but 6 perished in the disaster. The spiritual leader of Nigeria's Sunni Muslims, His Eminence, the Sultan of Sokoto, Alhaji Muhammadu Maccido was among the dead.

6

Global Aviation Industry: Analysing Share of Indian Aviation

CONTEMPORARY GLOBAL AVIATION INDUSTRY: AN OVERVIEW

In 2005, the global scheduled airline industry generated US$ 395 bn of revenues. After incurring huge aggregate losses for three consecutive years (2000–03), the world aviation industry has witnessed a strong growth in passenger traffic since 2004. However, there is still some time before airline companies start making profits. As per IATAs (International Air Travel Associations) latest estimates, the industry will book a smaller aggregate loss in 2006, before returning to modest profit in 2007.

The air travel market grew up originally to meet the demand of business travellers as companies became increasingly widespread in their operations. On the other hand, rising income levels and extra leisure time led holiday-makers to travel to faraway places for their vacation. A further stimulus to the air travel market was provided by the deregulation and the privatisation of the aviation industry. State-owned carriers that hitherto enjoyed monopoly status were now exposed to competition from private players. However, one development that changed the entire landscape of the industry was the emergence of Low Cost Carriers (LCCs). These carriers were able to offer significantly cheaper fares on account of their low-cost business models and thereby attract passengers who might not otherwise be willing to fly. LCCs have achieved rapid growth in market share in the U.S. domestic market, short-haul market in Europe and recently in Asia. Since 1970, the international passenger traffic has grown by an average rate of more than 6 per cent, compared to a 7 per cent increase in the domestic passenger traffic. The aviation industry is highly cyclical. However, in times of recession, the decline in the industry growth rate is much sharper when compared to the world economy. After witnessing a strong growth during the late 1990s, the industry saw a sharp reversal in fortune as a result of a global economic downturn in 2001. The situation was further aggravated by 9/11 attack, the Iraq war and the SARS epidemic. The mammoth financial losses incurred by the scheduled carriers during this period led to a long-overdue restructuring among the Full Service Carriers (FSCs). Many airlines embarked upon severe cost-cutting and fleet-rationalisation programmes as they struggled to remain afloat. The conditions for FSCs were further worsened with the advent of budget carriers in the U.S. and Europe. There was however a strong rebound in traffic in 2004, led by a strong recovery in the world economic growth and which continued for the next two years (2005–06). According to ICAO (International Civil Aviation Organisation), the revenue per passenger kilometers (calculated as the number of seats multiplied by the kilometers flown) for international services has grown by 8.5 per cent in 2005 and is estimated to have grown by 6 per cent in 2006. The strong growth in the traffic and recovery of higher fuel cost through surcharges resulted in strong revenue growth for airline companies. However, this did not translate into a recovery in profitability, primarily on account of a significant increase in fuel costs. According to IATA, the combined losses posted by the world's scheduled carriers amounted to US$ 6 bn in 2005, following a cumulative loss of US$ 36 bn in the previous four years. As per the estimates of aircraft manufacturers and other industry bodies, the world passenger traffic is expected to grow at 5 per cent p.a. in the medium to long-term. The growth will however be slower in matured economies, but faster in under-penetrated and growing economies like India and China. The primary reason for the increase in passenger traffic over the years has been decline in airline passenger yields. As per an estimate, after adjusting for the general inflation, the average airline yields (revenue per passenger kilometers) have almost halved since 1970. During the same period, the real revenue growth (by combining growth in traffic and decline in yields) has averaged only 2 per cent to 3 per cent. Since aviation industry is a high fixed cost industry, a small increase in operating cost can have a sharp impact on the profitability of the companies. High fuel prices, congestion cost, higher security and insurance cost can increase the overall cost of operations and thereby impact the demand for air travel services. However, there is room for cost reduction in the form of distribution cost and cost synergies from industry consolidation. Overall, we believe that consolidation is the only solution for addressing the problem of excess capacity and poor financial ratios of the company.

AIRLINE INDUSTRY PROFITABILITY IMPROVES: TURNING PEANUTS INTO SUSTAINABLE PROFITS BY IATA METHOD

The International Air Transport Association (IATA) announced a US$ 5 billion projected profitability for airlines in 2007 at their Annual General Meeting in Vancouver.

In his keynote State of the Industry address, Giovanni Bisignani, IATA Director-General and CEO said, "We expect airlines to make US$ 5 billion this year. The industry has changed tremendously in five years. Labour productivity is up 56 per cent. Distribution costs are down 13 per cent. Non-fuel unit costs dropped 15 per cent and load factors are at record highs—76 per cent in 2006. Airlines needed an oil price of less than US$ 20 per barrel in 2002 to break even. Today, we are profitable at nearly US$ 70 per barrel."

While the results are encouraging, Bisignani also gave an important reality check, "Airlines are a US$ 470 billion industry. A profit of US$ 5 billion is peanuts. We need US$ 40 billion just to cover the cost of capital. The industry is moving in the right direction, but with US$ 200 billion of debt, the financial hole is deep. The challenge for is to turn peanuts into sustainable profits."

IATA Outlined Six Priorities for the Industry

Safety

"In a decade we cut the accident rate in half. 2006 was our safest year ever, with 1 accident for every 1.5 million flights. And IATA member airlines did considerably better at 1 accident for every 2 million flights. The IATA Operational Safety Audit—a condition of IATA membership—is crucial to our efforts to make air travel even safer. Already 153 airlines are on the Registry and the industry is on target to improve safety a further 25 per cent by the end of 2008." Not all regions are performing at the same level. "While we are raising the bar on safety, IATA is working with its members to deliver results. IATA's Partnership for Safety has helped 89 airlines across Russia, Africa, the Middle East and Latin America."

Security

"Nearly six years after the tragic events of 2001, we are much more secure, but the system is a US$ 5.6 billion uncoordinated mess. Progress has been made on liquids and gels, but with more checked baggage, infrastructure issues must catch up. And governments have forgotten that our success on safety is based on global standards and cooperation. We need to take the same approach with security. It is a government responsibility to provide effective security that is also convenient. Our passengers have been hassled for six years. That's far too much," said Bisignani.

Simplifying the Business

"Simplifying the Business is a winner. In three short years, ET penetration jumped to 80 per cent. E-freight will be a reality in five locations by the end of this year. RFID will be part of a new approach to baggage handling. Forty-eight airlines are using bar coded boarding passes and CUSS is used in 59 airports around the world today. A US$ 6.5 billion revolution is taking shape and now we are looking over the horizon to stage two—a strategy for self-service that will define a new age for efficient travel," said Bisignani.

Infrastructure Costs

"Infrastructure providers must also share airlines' obsession with efficiency. While IATA achieved US$ 1.9 billion in cost savings from infrastructure partners in 2006, cost increases totalling US$ 2.6 billion wiped out the gains. While we are moving in the right direction with air navigation service providers, too many airports still operate as happy monopolies. Three airports—Bangkok, London and Paris—alone were responsible for US$ 1.4 billion of the US$ 2 billion of airport cost increases in 2006. Airlines live in a competitive world where commercial discipline is the regulator. Airport regulators are phantoms. Their financial results—with EBITs over 40 per cent—prove that it's a dream world for airports and a nightmare for airlines that pay the bill. The target for 2007 is to give the draft European Directive on airport charges some teeth by incorporating cost efficiency targets for the national regulators that it requires," said Bisignani.

Liberalisation

"This is the Everest challenge for the industry. Airlines compete in a modern world, but live in a time capsule with 60-year-old rules. It's like trying to climb Everest with equipment from 1944. The US-EU agreement is an important step in the right direction, opening up many new opportunities, but we lost a great chance for fundamental change. Governments must understand that we have modern businesses to run. Airlines need the freedom to sell their products where markets exist and to merge or consolidate where it makes business sense. Progressive liberalisation is the only way forward. The bilateral system belongs in a museum, in a display case next to the paper ticket," said Bisignani.

Environment

"The industry's toughest challenge is the environment. Our track record has limited airlines' carbon footprint to 2 per cent of the global total. By 2050, the UN estimates it will grow to 3 per cent. In today's political reality that is not acceptable—for any industry. We must aim high. Air transport must become an industry that does not pollute—zero emissions. We will not achieve this overnight, but potential building blocks for a carbon-free future are here. Turning this vision into reality will require leadership from governments, a global direction from the International Civil Aviation Organization (ICAO) and a technology leap from manufacturers and fuel suppliers. At the same time, IATA will work with airlines to implement best practice Environment Management Systems that place environmental responsibility alongside safety and security as a core promise to our 2 billion passengers a year.

"Industry leadership is critical. IATA has to set tough targets to meet the tough reality of business. But our biggest challenge is to have all of our members on board. So IATA has an important role in building the industry's leadership so that even small carriers can deliver great results," said Bisignani. IATA is strengthening the International Airline Training Fund (IATF) to help its member airlines on the industry's most critical issues: environment, safety and simplifying the business.

"Industry and government leaders must share a common view, addressing climate change with global solutions, tearing down the outdated bilateral system, ensuring cost efficient infrastructure, harmonising security and making travel convenient again. We are a great industry capable of innovation and change. So I am confident in our future: safe, secure, efficient and environmentally responsible," said Bisignani.

A REPORT ON THE PRESENT STATE OF THE AIR TRANSPORT INDUSTRY: AN IATA POINT OF VIEW AT AGN

Simplifying the Business

An IATA simplifying the Business is a winner. E-T is at 80 per cent. If you travelled here using a paper ticket, frame it and donate it to your local museum. It will soon be a part of our history. And to help make it happen, we have a team of 150 experts working directly with you. In the final lap, they will be working overtime to provide maximum support. Progress on the other projects is good. E-freight will be a reality in five locations this year, RFID will be part of a new approach for baggage handling, 48 airlines are using bar coded boarding passes and CUSS is being used in 59 airports. A US$ 6.5 billion revolution is taking shape and we are looking over the horizon to stage two, a strategy for self-service that will define a new age for travel.

Safety

Safety is another big success. 2006 was the safest year ever. In a decade, we cut the accident rate in half to 1 accident for every 1.5 million flights. Working together, IATA members did even better with 1 accident for every 2 million flights. Air is the safest way to travel and we are committed to make it even safer. Last year you made IOSA a condition of membership defining IATA as a quality association. 97.5 per cent of our members credible. Regrettably, we had to terminate 6 members that did not commit. Already 153 airlines are on the Registry, transparent for governments and passengers. While raising the bar on safety, we are working with our members to deliver results. The target is to lower the accident rate a further 25 per cent by 2008. We are strengthening our Partnership for Safety programme—that has helped 89 member airlines across Russia, Africa, Middle East and Latin America—to maintain support for our number one priority.

The Bottom Line

Your hard work is also improving your bottom line. Airlines nearly broke even in 2006 and we expect to make US$ 5 billion this year. Five years ago it was a different world. We needed an oil price of less than US$ 20 per barrel just to break even and low cost competition was new and tough.

You worked hard:

- Labour productivity: up 56 per cent,
- Distribution costs: down 13 per cent,
- Non-fuel unit costs: down 15 per cent, and
- Load factors: record high of 76 per cent.

Today we are profitable with oil at nearly US$ 70 per barrel and the low cost sector, 13 per cent of global operations, faces new challenges. For most, the short-haul carnival is over and long- haul will not be any easier. Moreover, the market is recognising your hard work and money is coming back into the industry. Airline stocks are up 50 per cent over last year and we are starting to see some consolidation. We could make a small celebration but hold the champagne: a reality-check is necessary.

Peanuts

We are now a US$ 470 billion industry and a profit of US$ 5 billion is peanuts. We need US$ 40 billion just to cover the cost of capital. We are moving in the right direction but the financial hole is deep. The industry is US$ 200 billion in debt. Our financial foundation is weak. Some economists predict another five years of strong growth but nobody can predict shocks and even a small one could have a big impact.

The economic boom could be a boomerang. The increase in plane orders is making efficient aircraft more expensive and harder to get. More importantly, labour pressure on wages could kill productivity gains. The bottom line is tough. You have worked hard and unions must understand that the happy days of the free lunch are gone forever. So, in a difficult environment how do we turn peanuts into sustainable profits? We need to get even more serious about change.

This year we have a tall agenda for you and for your association. The challenges are enormous like mountains. We have almost conquered some:

- Simplifying the Business,
- Efficiency, and
- Safety.

These are like the Rocky Mountains. Here in Vancouver, we are close enough to see the peaks. Others such as security and infrastructure costs, are well-known. But, like the Alps, they are far away and we cannot see their summits. And finally, the biggest challenges: liberalisation and environment—The Himalayas. Let's take one peak at a time.

Security first. This is the Matterhorn challenge; well-known to all but progress with governments is painfully slow. Nearly six years after the tragic events of 2001 we are much more secure but the system is still a US$ 5.6 billion uncoordinated mess. The common approach to liquids and gels is an improvement. But with more checked baggage, the infrastructure must also improve. And overall governments have failed. They have forgotten that our success in safety was based on global standards and cooperation. With security we must take the same approach. And where standards exist such as biometric identification, governments must be faster to implement. IATA is working with member airlines to implement Security Management Systems and we will keep shouting politely to remind governments of their responsibility to provide effective security that is also convenient. Our passengers have been hassled for six years. That's far too much.

Infrastructure Costs

The next peak is Mont Blanc, enormous but conquerable. The challenge is to make our infrastructure partners share our obsession for efficiency. Last year IATA achieved US$ 1.9 billion in cost savings. Unfortunately, you did not see it. Costs increased US$ 2.6 billion, destroying the gains. To be fair, we are moving in the right direction with ANSPs but too many airports still operate as happy monopolies. While you—their customers, struggle to reduce costs, airports added US$ 2 billion to your bill. Three are at the top of my special list: Bangkok, London and Paris. Combined, they grabbed US$ 1.4 billion. This is outrageous and I will continue to embarrass them with figures and facts. I personally battled a long list of airports with margins over 40 per cent: Sydney, Auckland, Hong Kong, Seoul and Johannesburg. I could go on but hands up: who in this room has reported a 40 per cent margin to their shareholders? I don't see many hands.

That's because airlines live in a competitive world and commercial discipline is our regulator. Airport regulators are phantoms. Look at the financial results of airports. Clearly, it's a dream world for them but a nightmare for airlines that pay the bill. Last year in Paris, Vice-President Barrot promised a directive on airport charges. He delivered a draft in January requiring national regulators across Europe. This is progress but it has no cost-efficiency targets. We do not need any more phantom regulators. We need a directive with real teeth.

Fundamental Change

Many battles have been fought since 2002. The Five Horsemen of the Apocalypse did not stop us. Our business is more efficient and delivering a small profit. We have a clear vision for safety, security and efficiency. Now I would like to focus on the Himalayan challenges that could change the future of our business: liberalisation and the environment.

Liberalisation

The Everest challenge is liberalisation. We compete in a modern world but we are living in a time capsule with sixty-year-old rules. We are trying to climb Everest with equipment from 1944. The US-Europe agreement is a step in the right direction creating many new opportunities. But we lost a great moment for fundamental change. Governments must understand that we have a modern business to run and we need the freedom to sell our products where markets exist and to merge or consolidate where it makes business sense. Let me be absolutely frank, liberalisation is the only way forward. The bilateral system belongs in a museum next to the paper ticket.

Environment

Another challenge is coming into view, the environment. Like the famous mountain K-2, it is potentially the toughest climb. Climate change is a real concern for our customers and a political priority for many governments. As the G-8 prepares to discuss environment later this week, let us clearly state our case.

Our track record is good, reducing noise 75 per cent and

improving fuel efficiency 70 per cent. The billions being invested in new aircraft will make airlines a further 25 per cent more fuel-efficient by 2020. So even as demand for travel increases, the increase in our carbon footprint will be limited. The UN estimates that we will grow from 2 per cent of total carbon emissions to just 3 per cent by 2050. But, we have been silent in our success and now we have a reputation crisis. That makes us an easy target for politicians who think green and see cash. Look at the UK. For the environment, Chancellor Brown doubled the Air Passenger Duty. We are paying a billion pounds for his green credentials. He helped the UK budget but what has he done for the environment? He hasn't planted any trees but he is lost in the woods.

Worse many government policies are schizophrenic. Some governments are promoting carbon offsets with no reliable way to measure the impact. And Europe is rushing to include aviation in emissions trading. But when it comes to real measures to help the environment, Europe is dragging its feet.

I was asked not to call the situation a complete mess, so I will be polite and call the actions of governments irresponsible and inconsistent. A Single European Sky could save up to 12 million tonnes of CO2 but instead of action we have a European circus. Fifteen years of talks, talks and nothing but talks.

My message to presidential candidates, Prime Ministers-in-waiting, and politicians everywhere is simple: we have had enough PR. It's time to deliver some real results.

IATA's environmental policy is clear and is aligned with the IPCC recommendations. First, invest in technology. Second, fly the planes effectively and third, make the inftrastructure efficient. And only then consider emissions trading, provided it is global, does not distort competition and is open with other industries.

Despite a good track record and a strong policy, we have been weak communicators. So we are working with you and our industry partners to raise the volume of our communications. But this is only part of the solution.

Today, I believe that the industry needs a wake-up call. Our carbon footprint is growing and that is not politically acceptable—for any industry. Climate change will limit our future until we change our approach from technical to strategic. Strategy starts with vision: For Toyota, it's an Earth Charter, GE has ecomagination and BP has Beyond Petroleum. So what is our vision?

Let me start the debate by aiming high. Air transport must become an industry that does not pollute.

Zero Emissions

I don't have all the answers but our industry started with a vision that we could fly. The Wright brothers turned that dream into reality and look at where we are now.

We can see potential building blocks for a carbon-free future. Fuel cell technology is here. The first solar powered aircraft is built and we can make fuel from biomass—today. The way forward is marked by four challenges. Let's start with the short-term.

The First Challenge is for Governments and ANSPs

To eliminate the 12 per cent inefficiency in air traffic management and cut this in half by 2012 to save 35 million tonnes of CO2 annually. Three mega-projects could deliver results quickly: an efficient Pearl River Delta in China, a next generation air traffic system for the US. The third is a specific challenge for Chancellor Merkel—to commit to a Single European Sky, real environmental progress for her G8 colleagues to follow.

The Second Challenge is for ICAO and its 190 Member States

Deliver a global emissions trading scheme. One that is fair, voluntary and effective. And to set demanding environmental targets for governments and industry. The relevance of ICAO depends on its ability to deliver.

In the Longer-term the Number one Challenge is Technology

Build a zero emissions aircraft in the next 50 years. The US, Europe, Canada, China, Brazil, Russia and Japan are the aerospace leaders. They must coordinate basic research and then compete to apply it effectively. To support this, we need a clean fuel but governments have cut funding for alternative fuel research. Meanwhile, oil companies are busy counting US$ 15 billion—what you paid in increased refinery margins last year. It's time for governments and the oil industry to make some serious investments so that in 10 years 10 per cent of our fuel is from alternative sources.

The Final Challenge is for Ourselves—The Airlines

To implement green strategies across our business your association is developing IATA Project Green. The goal is to help our members implement best practices that put environmental responsibility alongside safety and security. As a core promise to our 2 billion passengers, you have proven that we can deliver amazing results. In just 50 years we went from the Wright brothers to the jet age. So by working together with a common vision a green industry is absolutely achievable. And if some heavy lifting is needed, IATA is your sherpa.

Leadership

Our journey across mountains, points to the need for strong leadership. IATA has to set tough targets to meet the tough reality of our business. But our biggest satisfaction is having all our members on board so we have a role in building tomorrow's leaders. The International Airline Training Fund is an important tool, building skills, knowledge and leadership. To help smaller members deliver great results, the challenges will not get any easier. So we are enhancing the Training Fund, to support our members on the most important priorities:

- Safety,
- E-ticketing,
- Environment, and
- Leadership.

I am confident that our industry leadership is strong but at government level, there is a crisis. Traditional leaders, the US and Europe, have lost their way. Europe is micro-managing and the US cannot see beyond domestic politics. Meanwhile, the industry's centre of gravity is shifting East. China and India are the industry's stars and by 2010, Asia-Pacific will be the single largest market. The critical question will be: does Asia-Pacific have the institutions to play a global role?

Because leadership is not just about size, all regions need strong institutions to lead change.

For **Africa**, we must reduce the accident rate, work hard to retain labour skills, improve infrastructure and move forward with regional liberalisation.

For the **Middle East** the priority is managing growth effectively without hiding inefficiency and the US $ 38 billion of planned infrastructure investments must build a solid platform for the future.

For **Russia and the CIS**, safety is at the top of the agenda. Bringing business practices in line with global standards is a strong number two.

And in **Latin America**, government policy must meet with commercial reality. Regional liberalisation is a priority, we addressed the failed airport privatisation in Argentina but a long list of infrastructure issues remains. Challenges are everywhere so we need strong leaders everywhere.

Confident

Industry and government leaders must share a common view from the mountain tops, addressing climate change with global solutions, tearing down the outdated bilateral system, ensuring cost efficient infrastructure, harmonising security and making travel convenient again. We are a great industry capable of innovation and change. So I am confident in our future: safe, secure, efficient and environmentally responsible.

Global Aviation Industry at Belfast Conference

The future of the global aviation industry, which contributes over a trillion dollars to the worldwide economy every year, is to be discussed in Belfast in September 2007 at an event hosted by Queen's University Belfast, Invest NI and Bombardier, in conjunction with the American Institute of Aeronautics and Astronautics (AIAA).

The world's leading aviation researchers, designers, analysts, manufacturers, integrators and operators will be attending a series of key events under the banner 'Aviation: Industry without Borders'. The events will take place at the Europa Hotel, Belfast, from 18 to 20 September.

Identifying potential ways to address today's challenges in economics, the environment, new aircraft designs and the exploration of new systems analysis methodologies will be just some of the issues being discussed.

Speaking about the uniqueness of Aviation: Industry without Borders, Dr. Wilson Felder of America's Federal Aviation Association said: "This event will definitely be an exciting exchange of aviation and aviation system ideas in a unique international forum. This is certainly one of the very few times that such senior engineers from the United States, Europe and Asia will present their views on the future of aviation."

Professor Raghu Raghunathan from the School of Mechanical and Aerospace Engineering at Queen's University added: "For Belfast and Northern Ireland this is an exceptional recognition of our strong aviation heritage and will inspire both young and professional engineers alike."

Almost 200 such prospective young engineers have already signed up to take part in 'Flight Vehicles for the Future', a special session designed for 16-19 years old in collaboration with W5, Northern Ireland's interactive discovery centre. Mathematics and physics will be related to their practical applications in the area of aviation and the session will also include talks and interactive workshops focusing on innovations in the design and operation of futuristic vehicle concepts.

Other events taking place include an Invest NI sponsored opening reception at the Ulster Folk and Transport Museum on 18 September and the Bombardier sponsored AIAA evening awards banquet at the Europa Hotel on 19 September.

The main conference events will include the Seventh Aviation Technology, Integration, and Operations Conference (ATIO), the second Centre of Excellence for Integrated Aircraft Technology (CEIAT) International Conference on Innovation and Integration in Aerospace Sciences and the 17th Lighter-Than-Air Systems Technology Conference.

The conference general co-chairs are Dr. Wilson Felder, FAA, and Professor Raghu Raghunathan, CEIAT. Technical programme co-chairs are Dr. Danielle Soban, Georgia Institute of Technology, and Dr. Mark Price, Queen's University Belfast. Conference plenary speakers include Charles Leader, Next Generation Air Traffic System, FAA; Luc Tytgat, EUROCONTROL; Michael Friend, The Boeing Company; David Coughtrie, BAE Systems; Andrew Daw, BAE Systems; and John Green, Aircraft Research Association.

AVIATION INDUSTRY: BRIEF PROFILE OF EUROPEAN & GERMAN MARKETS

Executive Summary

The aviation industry note examines the structure of the global aviation industry and gives information about the European aviation industry and German aviation industry in particular. It discusses the changes sweeping the aviation industry, and the general economic conditions that influenced the industry, especially since the 1990s. The note also explores the negative impact of the US terrorist attacks on the industry and measures taken by the industry to overcome the same. Finally, the note discusses the problems faced by the German aviation industry and the measures taken by the industry to overcome those problems. The note is intended to act as supporting material for case studies pertaining to the global, European and German aviation industries and the companies operating in those industries.

The Global Scenario

The aviation industry can be broadly classified into military aviation and civil aviation. The term 'civil aviation' covers a activities except for those undertaken by the military and the government. According to the classification made by the International Air Transport Association (IATA), the aviation industry include components: civil aviation, cargo airlines, airports and travel agents.

GLOBAL AVIATION INDUSTRY GATHERS FOR KEY CHINESE AIR SHOW

The major players in the global aviation industry gathered on October 31, 2006—Vying for more of a slice of the booming Chinese market at the start of China's only international air show. The chiefs of civil and military aircraft manufacturers, parts suppliers and designers are among thousands of delegates at Airshow China 2006, in the southern boomtown of Zhuhai. Exhibitors include US plane maker Boeing, its European rival Airbus, French military giant Dassault Systems and Russia's Tupolev.

China's importance in the global aviation market was signalled last week when Airbus announced it would open an assembly plant in the northern city of Tianjin for its A320, which would eventually make four planes a month.

It also announced last week the purchase of 150 A320 planes in a deal worth about 10 billion dollars—bringing its total Airbus order to 300.

Speaking at the show, Airbus China president Laurence Barron said the event offered key access to the Chinese market.

"China is a very important market for us. Air traffic is doubling here every five years," Barron said. "It has already become a major market. We take China extremely seriously."

Airshow China is also a major showcase for military equipment.

State-owned weapons manufacturer China Aerospace Science and Industry Corporation (CASIC) is exhibiting a record number of missiles and fighter jets.

It is showcasing three ballistic missiles and modern Global Positioning System (GPS) guided bombs — satellite-guided long-range strike weapons.

The display demonstrates "the advanced capability and level of Chinese missile weapon systems to safeguard its territorial air, sea and land," CASIC said in a statement.

Andrei Chang, editor of security magazine Kanwa Defence Review, said China had shown its advanced technology in military hardware.

"It shows they have the capability to export the Western kind of this technology, that they can manufacture, they can design (weapons)," he said. "It shows that they have a lot of investment." One highlight of the show is China's most advanced fighter-pilot trainer the "Shanying", or Mountain Eagle. The two-seat, single-engine aircraft is capable of carrying 2,000-kilo (4,400-pound) weapon payloads such as short-range air-to-air missiles and rocket launchers and bombs. Daily demonstrations will be held at the show, which runs until November 5.

The third-generation L-15 advanced training aircraft is also a focus, although the much-anticipated FC-1 fighter aircraft, which China jointly developed with Pakistan, is not here despite expectations, with only a model on display.

Government aviation procurement officials were also among the delegates.

"In the past, this exhibition has proved vital for us," said Awzy el Hafiz, the Egyptian government's representative. The show, being attended by 550 exhibitors from 33 countries and regions, also features demonstration flights of new aircraft from Tupolev and Cessna, and a closing display by the Russian Air Force's Knights Aerobatic Team.

China's rapid economic growth is fuelling its mushrooming aviation market, which in turn is driving increased production of planes and engines. Boeing has forecast a 7.3 per cent annual growth rate in the Chinese commercial market, while Rolls-Royce, the maker of aircraft engines, believes it will need 2,600 new planes over a 20-year period.

AVIATION INDUSTRY CONSOLIDATION UNFOLDS $ 120BN OPPORTUNITY FOR INDIA

"As more passengers take to the skies, the race among airlines to gain as much market share in a fiercely competitive market that's plagued by high capital expenditure and increasing pressure on the revenue per seat, has led to the current run of Mergers & Acquisitions (M&As), making India the fastest consolidating market in the world with deals worth more than $600 million in less than three months."

"Such consolidation coupled with crashing fares brings with it the need to expand capacities, a fact that has also caught the fancy of global aviation suppliers queuing up to grab a share of the expected $ 120 billion order book for more than 1,100 aircrafts, associated infrastructure and support that will be needed over the next two decades." says Bundeep Singh Rangar, Chairman of IndusView, the India-focused cross-border advisory firm.

In the Indian aviation sector that's expected to grow at the rate of 20 per cent over the next five years, timing will be of essence. The sooner an operator can grow in to new markets and segments more efficient will be the revenue generation and profit realization via shared costs and resources. Expanding in to new markets faster is not easily possible by organic means leaving room only for inorganic expansion.

The most notable of deals, the $ 322 million purchase of Air Sahara by Jet Airways, India's largest full service private sector airline increases Jet's market share by 8 per cent to 33 per cent, a combined fleet of more than 80 aircraft and additional sought-after international routes.

On the other hand, the acquisition of Deccan Aviation, India's top low-cost carrier by Kingfisher Airlines, a subsidiary of UB Group, India's largest alcoholic spirits manufacturer makes the latter India's largest airline with a 34 per cent market share.

Much like Jet Airways has positioned Jetlite, earstwhile Air Sahara as a value carrier, Kingfisher plans to let Air Deccan remain an LCC, giving it a presence across passenger segments including low-cost, normal fare and premium services.

"Creating a large entity is one thing but making the venture profitable is another proposition altogether. Air France-KLM is one of the most profitable companies in Europe, in sharp contrast with American Airlines, the largest airline in the United States, which had difficulty in managing the merger of Trans World Airlines (TWA) with itself in April 2001." explains Rangar.

The Kingfisher-Air Deccan combine will help them bring operating costs down, lower expenses on spares, ground handling facilities, engineering, staff and crew—turn a saving of up to $73 million in the first year itself.

"As far as Jet Airways and Air Sahara merger is concerned, before even talking about turning a saving they still have to chalk out a time-line for a total integration. Jet should be prepared for considerable short-term losses as integration will be complex due to the fact that the two entities are culturally very different even though both are full service carriers."

After the acquisition, Jet Airways' renaming Air Sahara as Jetlite and positioning it as a value carrier, industry estimates suggest that Jetlite would need up to $ 200 million over the next two years to turnaround and absorb projected losses.

India's civil aviation industry has achieved the biggest ever growth in aircraft movement and passenger traffic in the last couple of years. Passenger traffic in the domestic airports increased by 22.3 per cent to 59.54 million, while aircraft movement increased by 14.2 per cent to 730,000.

There is a vast untapped potential in the form of the aspiring middle class which is also growing in affluence. We will most likely see the so-called no frills airlines turn a new leaf and the value proposition that they bring to the passenger will determine their success.

INDIAN AVIATION: A GLOBAL FLAVOUR

Despite forecasts about growing losses fuelling an imminent shakeout, Indian aviation industry continues to be the flavour of the season in the global arena.

The disappointing listing of Air Deccan and the continuous slide of the Jet Airways scrip has not affected the sentiment of global investors.

According to estimates by a top analyst at the International Air Transport Association (IATA), the investment needed by the Indian aviation industry is around $ 90 bn in the next 24 years and this provides big opportunities for investors world-wide.

The Centre for Asia Pacific Aviation (CAPA) is of the view that the pie could get even bigger with $ 40 bn investment potential in the next five to seven years.

Though estimates are diverse, there is consensus on the boom in India which could be turn out to be biggest the global aviation industry has seen so far. During the recent IATA conference in Paris, civil aviation minister Praful Patel and Air-India chairman-cum-managing director V. Thulasidas were the most sought after despite the industry's worries about high fuel prices.

In fact, Mr. Patel was the only civil aviation minister invited to address the global airline industry association.

While the civil aviation minister was urged to carry on with reforms, the Air-India chief was elected to IATA's board of governors. Mr. Thulasidas was also part of the high-profile meeting of the airline CEO forum, according to information reaching here from Paris.

IATA director-general Giovanni Bisignani's speech also contained several references to India's aviation boom and the need for more reforms to speed up growth.

Apart from the $ 30 bn that airlines of India are planning to spend on purchase of aircraft, the global industry's focus is on the huge investments that Indian airports would require to meet growing demand. Mr. Patel's mention about upgradation of 35 non-metro airports was watched keenly, industry sources said.

The implications of the proposed merger of Air-India and Indian were also discussed with interest. The mega merger would create a large airline which could take on global majors like Singapore Airlines and Emirates.

The ongoing merger of Air Sahara with Jet Airways was another Indian topic of interest at the IATA meet. "Considering the strong economic growth, investment potential could be far larger that what we are estimating now," said Mr. Kapil Kaul, chief executive officer of CAPA for the Indian sub-continent and west Asia.

India needs more plans and better-equipped airports to handle growing traffic and this calls for huge investments, he told ET. That Mr. Thulasidas rubbed shoulders with top executives representing Singapore Airlines, United Airlines and KLM at the CEO forum meeting shows the interest of the global industry in India, industry sources said.

The two-hour discussion of the CEOs touched upon various issues including investment and oil prices, industry sources said. Mr. Bisignani had visited India last year, indicating IATA's growing interest in the country's booming aviation sector.

The global airline industry body feels that India should allow foreign airlines to invest here and encourage growth of infrastructure through further reforms.

India Global Aviation Growth

The Indian aviation industry and its growth would be an integral driver of future global aviation outlook, claimed Giovani Bisignani, Director-General of the IATA (International Air Transport Association) at an interactive session on "Aviation in India: Great Opportunities and Great Challenges", organised by the Confederation of Indian Industry (CII), on Tuesday, October 18, 2007, New Delhi.

Bisignani stated that India had come up quite well in the field of civil aviation, making the expansion of air transport in the country amongst the fastest in the world. However, he added, that much remained to be done.

"If India has to compete with the challenge posed by China, it will have to improve its infrastructure at the airports in the country, most of which are not up to international standards."

Globally, airlines are a 400 billion dollar industry, which generates 1.3 trillion dollars in economic output. Stating that the potential of air transport was too great to be taken lightly, the Director-General of IATA asserted that India must undertake five immediate tasks to take full advantage of this potential.

Listing the priorities, he said that the focus should be on Enhanced safety, cost-effective improvement of infrastructure, reasonable taxation, allowing the airline operators to function as real business and effective use of technology to simplify the procedures.

Bisignani said that India's air transport infrastructure was out of date and steps were needed to improve the same. Not getting into the debate on airport privatization, he said that IATA was not interested as to who owned the airports, but it was concerned that they must be safe, cost-efficient and provide adequate capacity for further growth.

Another urgent concern of the IATA was the increasing taxation on aviation, not only in India but all over the globe. He said that he was happy that the Indian Government had abolished the fuel uplift levy in 2002, but at the same time disappointed that the US Dollar 36 million collected by the levy had not been returned to the airlines.

"Liberalisation is not something to be feared but to be anticipated" he stated. China has a clear plan to liberalise and SAARC countries were moving in the same direction.

Bisignani stated that the IATA was not setting targets only for governments but it was setting deadlines for itself too.

"We are committed to switch over completely to e-ticketing by December 2007 because we need the three billion dollars that will be saved, once e-ticketing is introduced. On this target date for e-ticketing, India is way behind at five per cent, as against a 40 per cent average globally," Bisignani said.

"Similarly e-freight too is a key initiative being pursued by IATA, with as many as 38 documents being required currently for shipments by air", he added.

"Each year the amount of paper wasted on freight documentation could be accommodated in 39 Boeing 747-400s or 81 A 300s", he said, giving the grim picture of the waste this entails.

Earlier, in his welcome address, V. Thulasidas, Chairman of the CII National Committee on Civil Aviation and Chairman and Managing Director, Air India stated that the airline industry had seen a strong growth momentum in the last couple of years in India.

"New airlines, acquisition of new aircraft fleet by existing players, private airlines joining India's national carriers on international routes, launch of low cost variants in the Indian skies and strong global interest in the Indian aviation sector, be it airlines or airport management had instilled this sense of confidence over the Indian skies", he stated.

Thulasidas said that CII believed that the momentum that the aviation industry was witnessing in India would also reflect in business tourism and overall economic growth in the country. At the same time he also pointed out the challenges that India faced, particularly the infrastructure challenge. Airline-industry maturity and Airport infrastructure had failed to keep pace with each other in India, he said.

There are also challenges on account of Air Traffic inefficiencies, with enormous wastage of fuel taking place at India's airports on account of these inefficiencies. An enormous wastage of fuel at India's airports and in the skies currently, comes at a time when airlines are already bearing the brunt of historical highs in fuel costs globally.

Stating that the Indian aviation industry was not functioning in isolation he pointed that the challenges before the Indian aviation industry was a reflection of the challenges that aviation faced globally.

Giving his concluding remarks, Dr. Vijay Mallaya, Member of the CII National Committee on Civil Aviation and Chairman and

MD of the Kingfisher Airlines thanked Bisignani for the strong agenda that IATA had taken up, on behalf of the Aviation Industry, globally.

INDIAN AVIATION: TAKES-OFF AND FURTHER CONSOLIDATION

The aviation sector in India has seen great traction in the last couple of years. Indian companies had placed orders worth more than $ 31 billion till last year and additional orders worth more than $ 3 billion were placed at the 45th Farnborough International Air-show in Farnborough, UK.

The recently concluded sixth Aero India 2007 air-show, Asia's premier biennial air show held in Bangalore at the Yelahanka Air Force Base highlighted the growing significance of the Indian Aviation market to global aviation companies and suppliers.

While the Aviation sector order book indicates the growing demand for air-travel in the country, the $ 322 million purchase of Air Sahara by Jet Airways, India's largest private sector airline, signifies the start of consolidation in the sector that's expected to grow 20 per cent over the next five years. India will need as many as 852 additional aircrafts over the next two decades worth more than $ 72 billion to meet the growing demand for air travel.

The merger will increase Jet's market share by 8 per cent for a total of 33 per cent. Jet will have a combined fleet of more than 80 aircraft and garner more landing rights and secure many sought-after international routes.

The Jet-Sahara deal, the largest so far in the Indian aviation sector, reminds of Air France-KLM that was created by the merger between French airline Air France and Netherlands-based KLM in May 2004. The merger created the largest airline company in the world in terms of total operating revenues, and the third largest in the world (the largest in Europe) in terms of passenger-kilometres.

Creating a large entity is one thing but making the venture profitable is another proposition altogether. Air France-KLM is one of the most profitable companies in Europe, in sharp contrast with American Airlines, the largest airline in the United States, which had difficulty in managing the merger of Trans World Airlines (TWA) with itself in April 2001.

Other deals in the Indian aviation sector in the making include the much awaited merger of the two state owned Indian and Indian

Airlines operating domestic and international services respectively apart from the proposed stake purchase by Kingfisher Airlines, owned by India's liquor baron Vijay Mallya in the largest low cost airline in the country, Air Deccan.

Some of the purchases by Indian carriers of aircrafts or engines during the year are listed in data below:

Airline	*Promotor*	*Aircraft/Engine*	*Order Size*	*Order Value ($ million)*
GoAir (Mumbai-based)	Wadia Group	Airbus A320	10	670
Flyington Freighters (Hyderabad-based)		Boeing 77F Cargo planes	4	945
Kingfisher Airlines (Mumbai-based)	Vijay Mallya	Pratt & Whitney 4000-100 engines	10	360
Air Deccan (Bangalore-based)	Deccan Aviation	IAE engines for A320	60	500

Chicago-based Boeing Company, the world's largest aerospace company, has confirmed orders valued at around $19 billion for 106 aircrafts of various configuration, including smaller and large wide-bodied planes, to be executed by 2011. The Boeing Company will invest $280 million in India for setting up an airplane maintenance facility and a pilot training centre.

GoAir, a startup Indian airline owned by Mumbai-based diversified Wadia Group, signed a purchase agreement with the European aircraft manufacturer Airbus at the Farnborough International Airshow to buy ten A320 planes for $670 million, with an option to buy further ten planes, taking the total order size to 20 planes for $ 1.3 billion.

Kingfisher a rapidly growing carrier in Indian civil aviation already flies 10 Airbus A320s and three Airbus A319s on domestic routes and will receive 30 more single aisle aircraft from Airbus in the near future.

Earlier, Kingfisher placed firm orders for five Airbus A380s, five A330s and five A350s for use on long-range international and regional services. So far, Kingfisher's Airbus Family fleet will grow to more than 60 aircraft.

Kingfisher Airlines had awarded a $24-million contract to Thales UK., the British unit of Thales SA, a leading electronic systems and industrial electronics company in defence, aerospace, airlines security and safety, for two A320 and one ATR full flight simulators during the air show.

Air Deccan, the first low-cost airline in India, signed a $ 17.5 million contract with Canada's CAE Inc., simulation and modelling technologies provider for civil aviation, to purchase an A320 full-flight simulator and a CAE Simfinity A320 Integrated Procedures Trainer.

This has happened as a result of the growing demand for air travel, falling prices of air-tickets and the emerging segment of low-cost airlines in India. During the financial year 2005-06, India's civil aviation industry achieved the biggest ever growth in aircraft movement and passenger traffic. Passenger traffic in the domestic airports increased by 22.3 per cent to 59.54 million, while aircraft movement increased by 14.2 per cent to 730,000.

While Indian aviation industry is emerging as a big buyer for the global vendors, the Indian IT industry is looking for a bigger share in the pie of aerospace technology outsourcing. HCL Technologies, the fifth largest IT services company of India, made their debut at Farnborough air-show this year. NASSCOM, the industry body representing about 900 Indian IT companies, estimates the market size of the aerospace technology outsourcing market in India to be about $150 million, and predicts $1 billion in the next four years.

India's Air Defence Allocations

India is not just an emerging commercial market, but also a big military market for the aerospace industry. It is expected to invite proposals for procuring and co-producing 126 multirole combat aircrafts. It is expected to include U.S. bids for F-18, F-16F aircraft in addition to traditional suppliers from Russia, France and the U.K. The deal size is estimated to be more than $11-billion, the biggest in more than a decade for fighter jet makers.

Fort Worth, Texas-based Bell Helicopter Textron, Inc is competing with the world's largest civil and military helicopter manufacturer Eurocopter to sell about 200 helicopters to the Indian army. Eurocopter, the wholly owned subsidiary of European aerospace company EADS (European Aeronautic Defence and Space), is also bidding for 260 single-engine high-altitude light reconnaissance helicopters for the Indian Army, and has plans to invest more than US$1 billion in India over the next two years.

Appendix 1

Aircraft Registration: Prefixes and Listings

Regn. Prefix	*Old Prefix*	*Country Name*	*Period*	*Remarks*
A-		Austria	1929-1939	Changed to OE-
AN-		Nicaragua	1936-	Changed to YN-
BR-		Burundi	1962-1965	Changed to 9U-
C-		Colombia	1929-1946	Changed to HK-
CB-		Bolivia	1929-1954	Changed to CP-
CCCP-		Soviet Union	1929-	Changed to RA-
CF-		Canada	1929-1974	Changed to C-
CH-		Switzerland	1929-1936	Changed to HB-
CR-A		Mozambique	1929-1975	Changed to C9-
CR-B		Mozambique	1971-1975	Changed to C9-
CR-C		Cape Verde Islands	1929-	Changed to D4-
CR-G		Portuguese Guinea/Guinea Bissau	1929-1975	Changed to J5-
CR-I		Portuguese India	1929-1961	Changed to VT-

(*Contd.*)

Regn. Prefix	*Old Prefix*	*Country Name*	*Period*	*Remarks*
CR-L		Angola	1929-1975	Changed to D2-
CR-M		Macau		Changed to CS-M
CR-S		Sao Tome and Principe	1929-	Changed to S9-
CR-T		Timor	1929-1975	Changed to PK-
CS-M	CR-M	Macau		Changed to B-M
CV-		Romania	1929-1936	Changed to YR-
CY-	VP-C	Ceylon	1948-1954	Changed to 4R-
CZ		Monaco	1929-1949	Changed to MC-
DDR-		East Germany	1945-1956	Changed to DM-
DM-		East Germany	1956-	Changed to D-
ES-		Estonia	1929-1939	Merged into Soviet Union CCCP-
EZ-		Saar Territory	1929-1933	Changed to SL-
FC-		Free French	1940-1944	Changed to F-
F-D		French Morocco	1929-1952	Changed to CN-
F-KH		Cambodia	1945-1954	Changed to XU-
F-L		Laos	1945-1954	Changed to XW-
F-O		Dahomey/Benin	1929-1960	Changed to TY-
F-O		Algeria	1929-1962	Changed to 7T-
F-O		Cameroon	1929-1960	Changed to TJ-
F-O		Congo	1929-1960	Changed to TN-
F-O		Gabon	1929-1960	Changed to TR-
F-O		Guinea	1929-1958	Changed to 3X-
F-O		Cote D'Ivoire	1929-1960	Changed to TU-
F-O		Madagascar	1929-1960	Changed to 5R-
F-O		Mali	1929-1960	Changed to TZ-

(*Contd.*)

Regn. Prefix	Old Prefix	Country Name	Period	Remarks
F-O		Mauritania	1929-1960	Changed to 5T-
F-O		Niger	1929-1960	Changed to 5U-
F-O		Senegal	1929-1960	Changed to 6V-
F-O		Tunisia	1929-1956	Changed to TS-
F-O		Ubangi-Shari	1929-1960	Changed to TL-
F-O		Upper Volta	1929-1960	Changed to XT-
F-O		Chad	1929-1960	Changed to TT-
F-VN		French Indochina (Vietnam)	1929-1954	Changed to XV-
J		Japan	1929-1945	
JZ-		Dutch East Indies	1945-	
K-		Kuwait	1967-1968	Interim-Changed to 9K-
KA-		Katanga	1961-1963	Unoffical-Changed to 9O-
KW-		Cambodia	1954	Changed to XU-
LG-		Guatemala	1936-1948	Changed to TG
LI-		Liberia	1929-1952	Changed to EL-
LR-		Lebanon	1944-1954	Changed to OD-
LY-		Lithuania	1929-1939	Merged into Soviet Union CCCP-
M-		Spain	1929-1933	Changed to EC-
MC-		Monaco	1949-1959	Changed to 3A-
OA-		Peru	1929-1950	Changed to OB-
OO-C		Belgium Congo	1929-1960	Changed to 9O-
PI-		Philippines	1941-1975	Changed to RP-
R-		Panama (two digits)	1929-1943	Changed to RX-
R-		Argentina (three digits)	1929-1932	Changed to LV-
RV-		Persia (Iran)	1929-1944	Changed to EP-

(*Contd.*)

Regn. Prefix	Old Prefix	Country Name	Period	Remarks
RX-		Panama	1943-1952	Changed to HP-
RY-		Lithuania	1929-1939	Merged into Soviet Union CCCP-
SL-	EZ-	Saar Territory	1947-1959	Changed to D-
SN-		Sudan	1929-1959	Changed to ST-
TJ-		Transjordania	1946-1954	Changed to JY-
TS-		Saar Territory	1930-1931	Unoffical-Changed to EZ-
UH-		Saudi Arabia	1945	Interim-Changed to HZ-
UL-		Luxembourg	1929-1939	Changed to LX-
UN-		Yugoslavia	1929-1935	Changed to YU-
VO-		Newfoundland	1929-1939	Merged into Canada CF-
VP-A		Gold Coast (Ghana)	1929-1957	Changed to 9G-
VP-B		Bahamas	1929-1975	Changed to C6-
VP-C		Ceylon	1929-1948	Changed to CY-
VP-G		British Guiana	1929-1967	Changed to 8R-
VP-H		British Honduras (Belize)	1947-	Changed to V3-
VP-J		Jamaica	1930-1964	Changed to 6Y-
VP-K		Kenya	1929-1963	Changed to 5Y-
VP-L		Leeward and Windward Islands	1929-	Now Antigua V2-
VP-LKA-LLZ		St. Kitts and Nevis		Changed to V4-
VP-M		Malta	1929-1968	Changed to 9H-
VP-N		Nyasaland	1929-1953	Changed to VP-Y
VP-P		Western Pacific Islands	1929-	Now Solomon Islands H4-
VP-R		Northern Rhodesia	1929-1953	Changed to VP-Y
VP-S		Somliland	1929-1960	Changes to 6OS-
VP-T		Trinidad and Tobago	1931-1965	Changed to 9Y-
VP-U		Uganda	1929-1962	Changed to 5X-

(*Contd.*)

Regn. Prefix	Old Prefix	Country Name	Period	Remarks
VP-V		St. Vincent and Grenadines	1959-	Changed to J8-
VP-W		China (Wei-Hai-Wei)	1929-1939	
VP-W		Rhodesia	1971-	Changed to Z-
VP-X		Gambia	1929-1945	
VP-Y		Southern Rhodesia, Rhodesia, Nyasaland	1929-1964	Changed to 7Q-, 9J-
VP-Z		Zanzibar	1929-1964	Changed to 5H-
VQ-B		Barbados	1952-1968	Changed to 8P-
VQ-C		Cyprus	1952-1960	Changed to 5B-
VQ-F		Fiji/Tonga/Friendly	1929-1971	Changed to DQ-
VQ-G		Grenada	1962-	Changed to J3-
VQ-L		St. Lucia	1965-	Changed to VQ-L
VQ-M		Mauritius	1929-1968	Changed to 3B-
VQ-P		Palestine	1930-1948	Changed to either TJ- or 4X-
VQ-S		Seychelles	1929-1977	Changed to S7-
VQ-ZA, -ZD		Basutoland	1929-1967	Changed to 7P-
VQ-ZE, -ZH		Bechuanaland	1929-1968	Changed to A2-
VQ-ZI		Swaziland	1929-1975	Changed to 3D-
VR-A		Aden	1939	
VR-B		Bermuda	1931-	Changed to VP-B
VR-C		Caymen Islands	1968-	Changed to VP-C
VR-G		Gibraltar	1929-1939	Changed to G-
VR-H		Hong Kong	1929-	Changed to B-H
VR-J		Johore	1929-1963	Changed to 9M-
VR-L		Sierra Leone	1929-1961	Changed to 9L-
VR-N		British Cameroons	1929-1958	Changed to either TJ-or 5N-
VR-O		Sabah (North Borneo)	1929-1963	Changed to 9M-
VR-R		Malaya	1929-1963	Changed to 9M-

(*Contd.*)

Regn. Prefix	Old Prefix	Country Name	Period	Remarks
VR-S		Singapore	1929-1965	Changed to 9V-
VR-U		Brunei	1929-	Changed to V8-
VR-W		Sarawak	1929-	
XH-		Honduras	1929-1960	Changed to HR-
XT-		China	1929-1949	
XV-	3W-	South Vietnam	1959-1975	
XW-		Laos	1954-	Changed to RDPL-
YE-		Yemen	1955-1969	Changed to 4W-
YL-		Latvia	1929-1939	Merged in Soviet Union CCCP-
YM-		Danzig Free State	1929-1939	Changed to D-
YN-		Nicaragua	1929-1936	Changed to AN-
ZM-		New Zealand	1929-1939	
3W-		Vietnam	1954-1959	Interim-Changed to XV-
4W-		Yemen	1969-	Merged into 7O-
6OS-		Somalia	1960-1969	Changed to 6O-
9O-		Zaire	1960-1966	Changed to 9Q-

Websites for knowing Aircraft Registration Prefixes Listings

- http://www.landings.com/_landings/pages/search/reg-prefix.html
- http://www.ivao.org/db/prefix/
- http://www.airportcitycodes.com/aaa/IRPDB.html
- daveg4otu.tripod.com
- http://www.airlinecodes.co.uk/regprefixhis.asp

Appendix 2

Airline Call-Signs

The flight crew of any airliner needs to talk to ground control or to air traffic control officers before, during, and after a flight. This tends to reduce the risks of their aeroplane bumping into other aeroplanes. A standard company call-sign will normally be spoken before the flight number by both the aircrew and the relevant flight controllers.

Most of these call-signs relate very closely to the airline's trade name. The digits making-up the flight number are shown in published time-tables of the airline and also on the arrivals and departures screens that are found at most airports (and on their websites). It is clearly important that the verbal instructions passed by air traffic control officers are heard and responded to in the specific aircraft that they are directing to turn/climb/decelerate, etc. For that reason, company call-signs are internationally regulated and agreed to be sufficiently distinctive. This helps to minimise the chances of misunderstandings, and thus the risks of aeroplanes leaving their authorised course.

This is a list of the less obvious call-signs in use by many airlines, in alphabetic order of airline name. A few air forces are included too:

Aircraft	*Call-sign*	*Country*
Aalborg Airtaxi	Dantex	Denmark
Aces—Aerolineas Centrales De Columbia	Aces	Columbia
Action Airlines	Action Air	USA
Aer Lingus	Shamrock	Eire
Aeris	Francecharter	France
Aero Aviation Centre	Sunrise	Canada
Aero Services	Bird Express	France
Aerolineas Argentinas	Argentina	Argentina
Aeromar Airlines	Trans Aeromar	Mexico

(Contd.)

Aircraft	*Call-sign*	*Country*
Aeropostale	Aeroposta	France
African International Airways	Flycargo	Swaziland
African Safari Airways	Zebra	Kenya
Air 2000	Jetset	UK
Air Alsie	Mermaid	Denmark
Air BC	Aircoach	Canada
Air Caledonie	Aircal	New Caledonia
Air Caledonie International	Aircalin	New Caledonia
Air Care	Drake	UK
Air Cargo America	Pegasus	USA
Air Cargo Carriers	Night Cargo	USA
Air Cat	Eurocat	Switzerland
Air Center West	Birdie	Denmark
Air Chaparral	Maverick	USA
Air Charter Services	Zaire Charter	Democratic Republic of Congo
Air Continental	Night Air	USA
Air Europa Express	Air Regional	Spain
Air Foyle	Broadspeed	UK
Air Foyle	Cutlass	UK
Air Foyle	Foyle	UK
Air Foyle	Kris	UK
Air Foyle/Sunjet	Skyman	UK
Air Holland Regional—In Receivership 1999	Orange	Holland
Air Jamaica	Juliet Mike	Jamaica
Air Liberia	November Lima	Liberia
Air Madagascar	Madair	Malagasay Republic
Air Mauritanie	Mike Romeo	Mauritania
Air Medical	Airmed	UK
Air Mongol/Miat	Mongol Air	Mongolia
Air Nelson	Mot Air	New Zealand
Air Newark	Newair	USA
Air Niagra Express	Databird	Canada
Air Nippon Kasushika Gaisha	Ank Air	Japan
Air Outre Mer/Minerve	Frenchlines	France
Air Panama International	Arpa	Panama
Air Portugal	Tap	Portugal
Air Schefferville	Scheff	Canada
Air Senegal	Sonatra Air Senegal	Senegal
Air Shot	Byte	UK
Air South Airways	Shawnee	USA
Air Tchad	Hotel Tango	Tchad
Air Traffic Gmbh	Snoopy	Germany
Air Vanuatu	Air Van	Vanuatu
Air Via	Varna Air	Bulgaria
Air Wales	Snowdon	UK
Air World	Envoy	UK
Airborne Express/Abx Air	Abex	USA

(*Contd.*)

Aircraft	*Call-sign*	*Country*
Airtours International	Kestrel	UK
Airtran Airways	Citrus	USA
Airways International	Airnat	USA
AJ Air Services	Ajay	UK
AJT Air International	Turjet	Russia
AMC Air Charter	Bavaria	Germany
America West Airlines	Cactus	USA
American Airlines Turboprops	Eagle Flight	USA
American Trans-Air	Amtram	USA
Arrow Air	Big A	USA
Atlantic Airways	Faroeline	Norway/Faroe Islands
Atlantic Coast Airlines	Blue Ridge	USA
Atlantic Island Air	Triangle	Iceland
Atlantic Southeast Airlines	Candler	USA
Atlas Air	Giant	USA
Aurigny Air Services	Ayline	UK
Aviaexpress	Avex	Hungary
Aviair	Caribou	Canada
Avial	Rushair	Russia
Bac Express Airlines	Rapex	UK
Bali Air	Biar	Indonesia
Bangladesh Biman Airlines	Bangladesh	Bangladesh
Baron Aviation Services	Show Me	USA
Bo-S-Aire Airlines	Bowman Air	USA
Bond Air Services	Ugabond	UK
Braathens Helikopter	Bee Copter	Norway
Brasil Central Lineas Aereos Regional	Central	Brazil
Britannia Airways Gmbh	Windsor	Germany
British Aerospace—Filton	Sapphire	UK
British Aerospace—Warton	Tarnish	UK
British Airways	Speedbird	UK
British European	Jersey	UK
British International Helicopters	Lion	UK
British Regional Airlines	British	UK
British West Indian Airways	West Indian	Trinidad and Tobago
Britten Norman	Vectis	UK
Buffalo Airways	Buffalo Air	USA
Business Air	Granite	UK
Business Aviation Services	Coastrider	Holland
Business Express	Biz Ex	USA
Cameroon Airlines	Cam-Air	Cameroon
Canada 3000 Airlines	Elite	Canada
Canadian Forces	Canforce	Canada
Cape Air	Cair	USA
Capital Air Service	Capair	USA
Capital Aviation	Prestige	UK
Caribbean Airways	Caribjet	Barbados
Centennial Airlines	Wyo Air	USA
Centre Airlines	Dutch Valley	USA
Centreline Air	Clifton	UK
Channel Express	Channex	UK

(Contd.)

Aircraft	*Call-sign*	*Country*
Chicago Express Airlines	Windy City	USA
China Airlines	Dynasty	Taiwan
China United Airlines	Lianhang	China
City Air (Chester)	Air Chester	UK
Cityjet	City Ireland	Eire
Civil Aviation Authority	Minair	UK
Cologne Commercial Flight	Tomcat	Germany
Continental Express	Jetlink	USA
Continental Micronesia	Air Mike	Guam
Cooper Aerial Survey	Surveyor	UK
Corporate Air	Air Spur	USA
Corse Mediterreanee	Corsemed	France
Coval Air	Odyssey	Canada
Dairo Air Services—Das Air Cargo	Dairair	Uganda
Danish Air Transport	Dantrans	Denmark
Delta Air Transport	Dat	Belgium
Deutsche Ba	Speedway	Germany
Dragonair	Loongair	China
Dravidian Aviation	Firestone	UK
Druk Air	Bhutan	Bhutan
Dynamic Air	Dynamite	Holland
Eagle Air Service	Swiss Eagle	Switzerland
East-West European	Silver Wing	Bulgaria
Eastern Air Executive	Eastex	UK
Eastern Airways	Eastflight	UK
Eastern Flying Service	Speedair	Canada
Easyjet Switzerland	Topswiss	Switzerland
Edinburgh Air Charter	Saltire	UK
Efs Flug Service Gmbh	Flightservice	Germany
Euro Exec Express	Echo Express	Sweden
Euro Flight Sweden	Eurowest	Sweden
Euroair Transport	Alexandra	?
Eurobelgian Airlines	Belstar	Belgium
European Air Taxi	Neat	UK
European Air Transport/Dhl	Eurotrans	Belgium
European Aviation Air Charter	Eurocharter	UK
Euroscot Express	Scotair	UK
Executive Aviation Services Ltd	Jester	UK
Executive Jet Aviation	Execjet	Switzerland
Express Airlines	Flagship	USA
Express One International	Longhorn	USA
Express One International	Spartus	USA
Farner Air Transport	Blue Strip	Hungary
Fast Air Carrier/Tnt	Faster	Chile
Faucett Peru	Charlie Foxtrot	Peru
Finnaviation	Bluebird	Finland
Finnish Air Force	Finforce	Finland
Flight Precision Ltd	Cobham	UK
Flugdienst Fehlhaber	Witchcraft	Germany
Flugfelag Austurlands	Austurflug	Iceland
Flugfelag Nordurlands	Norland	Iceland
Fr Aviation	Broadway	UK

(*Contd.*)

Aircraft	*Call-sign*	*Country*
Fr Aviation	Rushton	UK
French Air Force	Foxtrot Mike	France
French Air Force—Transport Sq	Cotam	France
Frontier Airlines	Frontier Flight	USA
Garuda Indonesia Airways	Indonesia	Indonesia
Gb Airways	Geebee	UK
German Air Rescue	Ambulance	Germany
Germania	Joker	Germany
Go Fly	Goflight	UK
Gojet Aviation	Gojet	UK
Gp Express Airlines	Regional Express	USA
Great Lakes Aviation	Lakes Air	USA
Great Wall Airlines	Changahang	China
Greenlandair	Glace	Denmark
Gt Aviation	Geetee Avia	UK
Gulfstream International Airlines	Gulf Flight	USA
Heliair	Heli-Bulgaria	Bulgaria
Helikopter Service	Helibus	Norway
Helikopterservice Euro Air	Swedcopter	Sweden
Hokkaido International Airlines	Air Do	Japan
Hummingbird Island Airways	Hum	Maldives
IMP Aviation Service	Bluenose	Canada
Indian Airlines	Indair	India
Indonesia Air Transport	Intra	Indonesia
Industries Air Charter	Induschart	France
Inter-Island Air	Ugly Van	USA
Irish Air Transport	Irish Trans	Eire
Island Air	Alakai	USA
Isles of Scilly Skybus	Scillonia	UK
Jamahiriya Libyan Arab Airlines	Libair	Libya
Japan Air Commuter	Commuter	Japan
Japan Air Lines	Japanair	Japan
Japan Asia Airways	Asia	Japan
Jetair Flight Academy	Speedfox	Denmark
Kd Air Corporation	Kay Dee	Canada
Kitty Hawk International	Arthur	USA
Knight Air	Knightway	UK
Kuwait Airways	Kuwaiti	Kuwait
Labrador Airways	Lab Air	Canada
Langtry Flying Group	Proflight	UK
Lansa—Lineas Aereas Nacionales	Uniform Lima	Honduras
Lapa—Lineas Aereas Privadas Argentinas	Lapa	Argentina
Leeds Central Helicopters	Pathfinder	UK
Legend Airlines	Legendary	USA
Liberia World Airlines	Hotel Yankee	Liberia
Libyan Arab Air Cargo	Libac	Libya
Lineas Aereas Canarias	LCN	Spain
Lineas Aereas Suramericanas	Suramericano	Columbia
Linjeflyg	Swedline	Sweden
Lionair	Lair	Luxembourg
London Executive Aviation	Lonnex	UK

(*Contd.*)

Aircraft	Call-sign	Country
Lte Lufttransport Espana	Funjet	Spain
Lufttransport Unternehmen	LTU	Germany
Luxaviation	Red Lion	Luxembourg
Macedonian Airlines	Macair	Greece
Macedonian Airlines	Makavio	Macedonia
Maersk Air	Bluestar	UK
Magadan Airlines	Mavial	Russia
Majestic Airlines	Magic Air	USA
Manx Airlines	British	UK
Markair	Interalas	USA
May Air Xpress	Beechnut	USA
Mcalpine Helicopters	Macline	UK
Meridiana	Merair	Italy
Merlin Express	Package Air	USA
Mesa Airlines	Air Shuttle	USA
Metropolitan Police	Special	UK
Miami Air International	Biscayne	USA
Middle East Airlines	Cedarjet	Lebanon
Midway Airlines	Jet Express	USA
Midwest Express Airlines	Skyship	USA
Million Air	Mill Air	USA
Mng Cargo Airlines	Black Sea	Turkey
Msr Flug Charter	Skyrider	Germany
Mtm Aviation	Aviatrans	Germany
Myanmar Airways	Unionair	Myanmar
Nantucket Airlines	Ack Air	USA
National Airlines	Red Rock	Not Known
National Airlines	Natchair	South Africa
Nawa Air Transport	Supernawa	Hungary
New York Helicopter	Metrocopter	USA
Newair Airservice	Newdan	Denmark
Newfoundland And Labrador Air Transport	Nalair	Canada
Nigel Mansell	Flight Red	UK
Norsk Air	Norseman	Norway
North British Airlines	Teesair	UK
North Coast Aviation	Air Ohio	USA
Northamptonshire School Of Flying	Norton	UK
Northern Air Cargo	Yukon	USA
Northern Air Service	Nortax	USA
Northern Executive Aviation	Neatax	UK
Novair/Nova Airlines	Navigator	Sweden
Nurnbergen Flug	Flamingo	Germany
Nwt Air/Air Canada Connector	Territorial	Canada
Oman Air	Oryx	United Arab Emirates
Ostfriesische Lufttransport	Ostfries	Germany
Pan American World Airways	Clipper	USA
Panorama Air Tours	Lani	USA
Pegasus Airlines	Sunturk	Turkey
PLM Dollar Group	Osprey	UK
Polo Aviation	Chukka	UK
Polynesian Airways	Polyair	USA

(*Contd.*)

Aircraft	Call-sign	Country
Prince Edward Air	Comet	Canada
Pro Air	Prohawk	USA
Propheter Aviation	Air Prop	USA
Provincial Airlines	Starlite	Canada
Qinetiq, Boscombe Down	Gauntlet	UK
Quick Air	Dagobert	Germany
Ravenair	Raven	UK
Raytheon Corporate Jets	Manta	UK
Rex Aviation	Tranzair	New Zealand
Romavia	Aeromavia	Romania
Ross Aviation	Energy	USA
Royal Air Force	Rafair	UK
Royal Air Force—33 Group	Vortex	UK
Royal Air Force Air Support Command	Ascot	UK
Royal Air Maroc	Maroc Air	Morocco
Royal Airlines	Roy	Canada
Royal Australian Air Force	Aussie	Australia
Royal Flight	Kitty/Kittyhawk	UK
Royal Flight Helicopters (UK)	Rainbow	UK
Royal Navy	Navy	UK
Royal New Zealand Air Force	Kiwi	New Zealand
Royal Swedish Air Force	Swedic	Sweden
Ryanair	Budget Jet	Eire
Saber Cargo Airlines	Freighter	USA
Safair	Cargo	South Africa
Sata International	Air Acores	Portugal
Scot Airways	Suckling	UK
Servair	Swissbird	Switzerland
Sichuan Airlines	Chuabhang	China
Sierra National Airlines	Selair	Sierra Leone
Skyfreighters	Skybus	USA
Skyservice	Skytour	Canada
Skyways Regional Ex-Airborne of Sweden	Midwing	Sweden
Skywings	Skyfox	Switzerland
SMAC—Sabang Merauke Air Charter	Samer	Indonesia
SMB Stage Line	Stage	USA
South African Airways	Springbok	South Africa
South East Police Aviation Unit	Police	UK
South West Air	Saint Clair	Canada
Southwest Air Lines	Nansei	Japan
Spanair	Sunwing	Spain
Spanish Air Force	Air Mil	Spain
Spirit Airlines	Spirit Wings	USA
Sports Air Travel	West Coast	USA
Springdale Air Service	Spring Air	USA
Star Air	White Star	Denmark
Star Europe Airlines	Starway	France
Stratford Airways	Festival City	Canada
Suburban Air Freight	Sub Air	USA
Sun Country Airlines	Sunshine	USA
Sun-Air	Sunscan	Denmark
Sunaire	Sunbird	USA

(Contd.)

Aircraft	*Call-sign*	*Country*
Sunways Intersun Havacilik	Intersun	Turkey
Sunwest Airlines	Sunflight	Canada
Superior Aviation	Spend Air	USA
Surrey And Kent Flying Club	Surrey	UK
Swedewings	Wings	Sweden
Swiftair	Swift	Spain
Swiss Air Ambulance	Swiss Ambulance	Switzerland
Systems International Airways	Florida	USA
TEA Switzerland	Transway	Switzerland
TACV-Transportes Aereos De Cabo Verde	Transverde	Cape Verde Islands
Teddy Air	Teddys	Norway
Thy—Turkish Airlines	Turkair	Turkey
Thyssen Krupp Ag	Blueline	Germany
Titan Airways	Zap	UK
TNT International Avn	Nitro	UK
Tower Air	Tee Air	USA
Trans Air Link	Sky Truck	USA
Trans Continental Airlines	Transcon	USA
Trans European Airlines	Eurolines	Greece
Trans Mediterranean Airways	Tango Lima	Lebanon
Trans States Airlines	Water Ski	USA
Trans World Express	Redbird	USA
Transaero Airlines	Tango Oscar	Russia
Transporte E Trabalho Aero	Transaero	Mozambique
Transportes Aereos Neuquen	Taneu	Argentina
Tropic Air	Tropiser	Antigua
Tulip Air Charter	Farnair Europe	Holland
Tunis Air	Tunair	Tunisia
UK Royal Flight	Kitty or Kittyhawk	UK
United Air Services	Unitair	Zimbabwe
United Parcel Service	UPS	USA
Universal Airlines	Airgo	USA
US Airways	US Air or US Air Shuttle	USA
US Check Airlines	Star Check	USA
USA Jet Airways	Jet USA	USA
Uzbekistan Airways	Uzbek	Uzbekistan
Vanguard Airlines	Vanguard Air	USA
Viking Express	Titan Air	USA
Virgin Air	Virair	USA
Virgin Express France	Ladybird	France
Virgin Express Ireland	Green Isle	Eire
Virgin Islands Seaplane Shuttle	Waterbird	USA
VLM	Rubens	Belgium
Volga-Dnepr	Volga	Russia
Waglisla Air	Sea Hawk	Canada
Warbelows Air Ventures	Ventaire	USA
Warwickshire Aerocentre	Airtax	UK
West Air Sweden	Air Sweden	Sweden
West Coast Air	Gambia Coastal	Gambia

(*Contd.*)

Aircraft	*Call-sign*	*Country*
Western Pacific Air Service	Westpac	Solomon Islands
Williams Racing	Grandprix	UK
Woodgate Air Services	Woodair	UK
Worldways Canada	Canada	Canada
Young Flying Service	Young Air	USA

Some of the operators listed may have ceased trading or merged or updated their names. The country stated is the one in which the majority of their aircraft are believed to be registered, it may not always be the country in which the operator is legally incorporated. Airlines whose names are self-evident from the similarity of their call-signs are generally omitted from this list.

Appendix 3

International Air Transport Association (IATA)

IATA is an international trade body, created some 60 years ago by a group of airlines. Today, IATA represents over 240 airlines comprising 94 per cent of the international scheduled air traffic. The organisation also represents, leads and serves the airline industry in general.

Fact Sheet: International Air Transport Association (IATA)

Mission

- To represent, lead and serve the airline industry.

Leadership

- IATA Director-General and CEO, Mr. Giovanni Bisignani.
- Chairman of IATA Board of Governors (2007-08), Fernando Pinto, CEO, TAP Portugal.

Membership

- IATA membership: 243 airlines, 130 countries.
- Flights by IATA members represent 94 per cent of world's international scheduled traffic (Available Seat Kilometres).
- IATA members total freight—42.3 million tonnes in 2006, of which 27.5 million tonnes were international.
- IATA members total passengers 2006 (scheduled)—1.58 billion of which 655 million were international.

History

- IATA was founded in Havana, Cuba, in April 1945.

Offices

- IATA has 84 offices in 73 countries covering 115 nations.

- Headquarters: Montreal, Canada.
- Executive Office: Geneva, Switzerland.
- Regional Offices in Amman, Beijing, Johannesburg, Miami, Singapore.

IATA Annual General Meeting and World Air Transport Summit

Formalises industry positions on industry and public policy issues and provides a focus for emerging industry issues attended by representatives from IATA member airlines, industry partners, international and regional associations, manufacturers and suppliers:

- The first AGM was held in Montreal, Canada in October 1945.
- The 2007 AGM was held in Vancouver, Canada.
- The 2008 AGM will be held in Istanbul, Turkey on 1-3 June, 2008.

IATA Financial Services

IATA handled US$ 269 billion for airlines in 2006 through its financial systems.

- IATA Clearing House—US$ 40 billion.
- IATA Currency Clearing Service — US$ 22 billion.
- Billing and Settlement Plan (BSP) — US$ 187 billion.
- Cargo Account Settlement Systems (CASS) — US$ 20 billion.

2007 Priorities

Safety

95 per cent of Member airlines IOSA to be audited by 31 December, 2007.

Simplifying the Business

- 100 per cent e-ticketing in IATA BSPs by 31 May, 2008.
- 80 airlines have the capability to issue 2D bar-code enabled boarding passes by end 2007.
- 70 airports to offer Common Use Self-Service check-in (CUSS) facilities by end 2007.
- 5 e-freight pilot projects to be launched and completed.

Financial

- Save US$ 1.5 billion in taxes, charges and fuel fees,
- Including US$ 800 million in real cost reductions,
- Secure draft EC Directive on airport charges that meets key IATA requirements,
- Launch 8 new BSPs/CASSs, and
- Launch 4 new cost-saving and service improvement initiatives.

Operations

- Save US$ 1.5 billion through operational efficiencies, new routes and infrastructure enhancements,
- Harmonise Passenger Name Record (PNR) requirements; measure and improve effectiveness of passenger security processes.

Environment

- Ensure ICAO Assembly support for industry positions on climate change.
- Launch a major communications campaign, initially in Europe, to strengthen aviation's public position.

IATA at the Air Transport Industry's Side

Air transport is one of the most dynamic industries in the world. The International Air Transport Association (IATA) is its global trade organisation.

Over 60 years, IATA has developed the commercial standards that built a global industry. Today, IATA's mission is to **represent, lead and serve the airline industry**. Its members comprise over 240 airlines—the world's leading passenger and cargo airlines among them—representing 94 per cent of international scheduled air traffic.

Representing

IATA seeks to improve understanding of the industry among decision makers and increase awareness of the benefits that aviation brings to national and global economies. It fights for the interests of airlines across the globe, challenging unreasonable rules and charges, holding regulators and governments to account, and striving for sensible regulation.

Leading

IATA's aim is to help airlines help themselves by simplifying processes and increasing passenger convenience while reducing costs and improving efficiency. The groundbreaking *Simplifying the Business* initiative is crucial in this area. Moreover, safety is IATA's number one priority, and IATA's goal is to continually improve safety standards, notably through IATA's *Operational Safety Audit* (IOSA). Another main concern is to minimise the impact of air transport on environment.

Serving

IATA ensures that people and goods can move around the global airline network as easily as if they were on a single airline in a single country. In addition, it provides essential professional support to all industry stakeholders with a wide range of products and expert services, such as publications, training and consulting. IATA's financial systems also help carriers and the travel industry maximise revenues.

For the benefit for all parties involved:

- For *consumers,* IATA simplifies the travel and shipping processes, while keeping costs down. Passengers can make one telephone call to reserve a ticket, pay in one currency and then use the ticket on several airlines in several countries.
- IATA allows *airlines* to operate safely, securely, efficiently and economically under clearly defined rules.
- IATA serves as an intermediary between airlines and *passenger* as well as *cargo agents* via neutrally applied agency service standards and centralised financial systems.
- A large network of *industry suppliers* and *service providers* gathered by IATA provides solid expertise to airlines in a variety of industry solutions.
- For *governments,* IATA seeks to ensure they are well-informed about the complexities of the aviation industry to ensure better, long-term decisions.

IATA's Industry Priorities for 2007

Safety

- 95 per cent of Member airlines IOSA audited by 31 December, 2007.

Simplifying the Business

- 100 per cent E-ticketing in IATA BSPs by end May, 2008.
- 80 airlines 2D bar-code enabled by end 2007.
- 70 airports to offer CUSS facilities by end 2007.
- 5 e-freight pilot projects to be launched and completed.

Financial

- Save US$ 1.5 billion in taxes, charges and fuel fees (US$ 800 million real cost reductions).
- Secure draft EC Directive on airport charges that meets key IATA requirements.
- Launch 8 new BSPs/CASSs.
- Launch 4 new cost-saving and service improvement initiatives.

Operations

- Save US$ 1.5 billion through operational efficiencies, new routes and infrastructure enhancements.
- Harmonise Passenger Name Record (PNR) requirements; measure and improve effectiveness of passenger security processes.

Environment

- Ensure ICAO Assembly support for industry positions on climate change.
- Launch a major communications campaign, initially in Europe, to strengthen aviation's public position.

Gateway for Aviation Intelligence and Analysis (GAIA)

To lead an aviation industry, you need as much information as you can get. GAIA is IATA's new one-stop online source for market intelligence, statistics and trends in the aviation industry.

It includes five complete modules for different sectors of the industry:

Passenger Data

- Monthly International Statistics (MIS) Module.
- World Air Transport Statistics (WATS) Module.
- Passenger Forecast Module.

Cargo Data

- Freight Forecast Module.

Financial Data

- Airline Economic Results and Prospects.

This new application has the following benefits:

- Easy to navigate and search functions.
- Instant update of latest statistical data, GAIA is linked directly to our IATA central statistical database.
- Save all your custom queries.
- All data can be extracted into multiple formats.

Having access to comprehensive business intelligence allows you to spend your valuable time on data analysis rather than data compilation. In addition, GAIA provides:

- The basis for *planning and development* for all sectors of the aviation industry, associated industries and suppliers.
- A solid foundation for performing a realistic assessment of costs, which *reduces the risk of unwise investment* and helps maximise the return on this investment.
- Unique access to information that comes directly *from IATA's Member airlines*, whenever and wherever you need it.
- The basis for cutting-edge benchmarking and trend analysis enabling you to also *strengthen your forecasting accuracy*.

Updated throughout the year, except for Monthly International Statistics, updated monthly.

Our pricing policy depends on the number of user licenses and type of module.

IATA Partners

IATA Strategic Partners have been contributing to IATA through the Strategic Partnerships since 1990. They now number more than 300, supplying expertise across a wide range of activities and providing support to IATA and its Member airlines in a growing number of initiatives.

IATA's Corporate Structure

IATA Consists of the Office of the Director-General:

Director-General and Chief Executive Officer	*Giovanni Bisignani*
Human Capital	*Guido Gianasso*
Legal Services	*Robert McGeorge*
Information Technology Services	*Sam Sahana*
Audit Services	*Frank Di Stefano*
Communications	*Anthony Concil*
Chief Economist	*Brian Pearce*

Plus Five Major Divisions:

Corporate Services	*Bob Hutt*
Industry Distribution and Financial Services	*Tom Murphy*
Marketing and Commercial Services	*Mark Hubble*
Member and Government Relations & Corporate Secretary	*Thomas Windmuller*
Safety, Operations and Infrastructure	*Günther Matschnigg*

Giovanni Bisignani
Director-General and CEO

Mr. Bisignani has served as Director-General and CEO of the International Air Transport Association (IATA) since June 2002. During his tenure with IATA he has completely restructured the association to increase its relevance and speed in driving a broad agenda for industry change.

Since 2002 IATA has spearheaded an industry agenda to enhance safety, improve efficiency and focus governments on the long-term issues for industry viability—including liberalisation and a common-sense approach to issues such as the environment and taxation. As a result, two major IATA initiatives are re-shaping important parts of the industry. The IATA Operational Safety Audit (IOSA) is raising the bar on safety with the first global standard for airline operational safety management. And IATA's simplifying the business programme will use technology to improve customer convenience while eliminating US$6.5 billion in annual costs.

Mr. Bisignani's airline experience includes 5 years at the helm of Alitalia as CEO and Managing-Director, during which time he also served on the IATA Executive Committee and was Chairman of the AEA (Association of European Airlines). Prior to joining IATA, he launched and directed Opodo—the first European airline-owned online travel agency. He has been a Member of Pratt and

Whitney Advisory Board and Chairman of the global distribution services provider Galileo International.

During his business career, Mr. Bisignani has held various high-level responsibilities at the energy company ENI and the Italian industrial conglomerate IRI Group. He served as President of Tirrenia di Navigazione, the largest Italian ferry company and as CEO and Managing-Director of SM Logistics, a group of logistics and freight forwarding companies, partially-owned by GE Capital.

IATA's Corporate Governance Structure

This section contains the formal reference materials covering the IATA corporate governance structure.

Act of Incorporation

On December 18, 1945 assent was given to a special Act of the Canadian Parliament incorporating IATA.

Articles of Association

The Articles of Association regulate the activities and affairs of IATA. The Articles are amended from time-to-time by the Annual General Meeting. The most recent amendments were adopted by the 61st Annual General Meeting held in Tokyo 29-31 May, 2005.

The Nominating Committee

The Nominating Committee is appointed to make recommendations to the Annual General Meeting for appointments of members to the Board of Governors.

Rules of Procedure of General Meetings

The Rules of Procedure of General Meetings sets the composition, agenda, conduct of business, etc. for IATA's Annual General Meetings.

Board of Governors

The IATA Board of Governors is IATA's government. Its members exercise an oversight and executive role on behalf of the membership as a whole in representing the interests of the Association.

Industry Committees

Rules and Regulations of the six Industry Committees: Cargo, Environment, Financial, Industry Affairs, Legal,. And Operations.

Industry Committees Mandates and Membership

Working Groups Mandates and Membership

- Cargo,
- Environment,
- Financial,
- Industry Affairs,
- Legal, and
- Operations.

IATA Human Capital

At IATA, **we are putting our people at the centre of the business agenda.** Over the past few years, the air transport industry has met several challenges. In order to ensure its success, it had to simplify the way it does business.

In that context, IATA is looking for people who:

- Can **adapt quickly** to a changing environment.
- Are **team players.**
- Are **results-oriented.**
- **Value excellence** in customer-service.

At IATA, we are working hard at providing our employees with a great working and learning environment. Here are some of our **Human Capital priorities for 2007**:

- Reviewing our organisational structure creatively and professionally.
- Attracting, selecting and retaining the best employees
- Helping our employees go through changes by increasing our focus on management development and improving our internal communications.
- Maintaining a results-driven corporate culture by directly measuring achievements against accountabilities and behaviours.

As some of our Members and partners, we have gone through an extensive re-engineering of the ways we manage our staff.

In 2003, we introduced new staff selection process, performance assessment and rewards systems.

In 2004, we improved our structure to make IATA a **well-integrated organisation** by introducing the matrix model. Our goal was to facilitate cross-divisional collaboration and circulation of best practices.

In 2005, we reviewed our organisational design to **reduce the number of layers** and broaden the span of control. Additionally, we developed a **dynamic sales organisation** and culture.

In 2006, we consolidated European regional offices into one unique hub and drove staff development from simply training to focus on a leadership and learning role.

Also, in cooperation with Dr. Noel Tichy of the University of Michigan School of Business, we have successfully launched the IATA Change Leadership (ICL) and the IATA Leadership Development Programme (ILDP), which act as incubators for preparing the next generation of IATA leaders.

Additionally, the IATA Internship Programmes (IIP) allows us to attract and develop—on a worldwide basis—the best among young talents interested in joining the air transport industry.

History

IATA—The International Air Transport Association—was founded in Havana, Cuba, in April 1945. It is the prime vehicle for inter-airline cooperation in promoting safe, reliable, secure and economical air services—for the benefit of the world's consumers. The international scheduled air transport industry is now more than 100 times larger than it was in 1945. Few industries can match the dynamism of that growth, which would have been much less spectacular without the standards, practices and procedures developed within IATA.

At its founding, IATA had 57 Members from 31 nations, mostly in Europe and North America. Today it has over 270 Members from more than 140 nations in every part of the globe.

The modern IATA is the successor to the International Air Traffic Association founded in the Hague in 1919—the year of the world's first international scheduled services.

The old IATA was able to start small and grow gradually. It was also limited to a European dimension until 1939 when Pan American joined. The post-1945 IATA immediately had to handle worldwide responsibilities with a more systematic organisation and a larger infrastructure.

This was reflected in the 1945 Articles of Association and a much more precise definition of IATA's aims than had existed before 1939.

- To promote safe, regular and economical air transport for the benefit of the peoples of the world, to foster air commerce, and to study the problems connected therewith;

- To provide means for collaboration among the air transport enterprises engaged directly or indirectly in international air transport service; and
- To cooperate with the newly created International Civil Aviation Organization (ICAO—the specialised United Nations agency for civil aviation) and other international organisations.

The most important tasks of IATA during its earliest days were technical, because safety and reliability are fundamental to airline operations. These require the highest standards in air navigation, airport infrastructure and flight operations. The IATA airlines provided vital input to the work of ICAO, as that organisation drafted its Standards and commended Practices. By 1949, the drafting process was largely complete and reflected in "Annexes" to the Chicago convention, the treaty which still governs the conduct of international civil aviation.

In those early days, ICAO coordinated regional air navigation and support for airports and operational aids in countries which could not themselves afford such services. IATA provided airline input to ICAO and to sessions of the International Telecommunications Union on wavelength allocation.

The standardisation of documentation and procedures for the smooth functioning of the world air transport network also required a sound legal basis. IATA helped to mesh international conventions, developed through ICAO, with US air transport law which had developed in isolation prior to World War Two. The Association made a vital input to the development of Conditions of Carriage the contract between the customer and the transporting airline. One early item on the legal agenda was revision and modernisation of the Warsaw Convention—originally signed in 1929—on airline liability for passenger injury or death and cargo damage or loss. This work continues.

Once they were operating within a sound technical and legal framework, airlines' next requirements were for answers to questions such as: who can fly where? What prices are to be charged? How is the money from multi-airline journeys—that is, interlining—to be divided up, and how do airlines settle their accounts?

The Chicago Conference of 1944 which gave birth to the Chicago Convention tried to achieve a multilateral answer to the first-two questions, but failed to do so. The questions of who flies, and where, were resolved on a bilateral basis. The benchmark Bermuda Agreement of 1946 between the US and the UK was the first of almost 4,000 bilateral air transport agreements so far signed and registered with ICAO.

In the early days, governments insisted on the right to oversee the prices charged by international airlines but could not, in practical terms, develop those prices for themselves. IATA was delegated to hold Traffic Conferences for this purpose, with all fares and rates subject to final government approval. The aim was two-fold: ensuring that fares and rates would not involve cut-throat competition, while ensuring that they could be set as low as possible, in the interests of consumers.

A coherent pattern of fares and rates pattern was established, avoiding inconsistencies between tariffs affecting neighbouring countries—and thereby avoiding traffic diversion. The predictability of fares and rates in this pattern also enabled airlines to accept each others' tickets on multi-sector journeys and thus gave birth to interlining. Today, 50 million international air passengers a year pay for their ticket in one place, in one currency, but complete their journey using at least two, and sometimes five or more, airlines from different countries using different currencies.

The first worldwide Traffic Conference was held in Rio de Janeiro in 1947. It reached unanimous agreement on nearly 400 resolutions covering all aspects of air travel.

Fare construction rules for multi-sector trips, revenue allocation—pro-rating—rules, baggage allowances, ticket and air waybill design and agency appointment procedures were typical details agreed at this pioneering meeting.

Today, that pioneering work is reflected in the currently applicable IATA Resolutions dealing with these and many other subjects. Notable examples are:

- *The Multi-lateral Interline Traffic Agreements*: These are the basis for the airlines' interline network. Close to 300 airlines have signed them, accepting each others' tickets and air waybills—and thus their passenger and cargo traffic—on a reciprocal basis.

- *Passenger and Cargo Services Conference Resolutions*: These prescribe a variety of standard formats and technical specifications for tickets and air paybills.
- *Passenger and Cargo Agency Agreements and Sales Agency Rules*: These govern the relationships between IATA Member airlines and their accredited agents with regard to passenger and cargo.

Debt Settlement between airlines, largely arising from interlining, takes place through the Clearing House, which began operations in January 1947. During its first year, 17 airlines cleared (US) $ 26 million.

International air transport grew at double-digit rates from its earliest post-1945 days until the first oil crisis in 1973. Much of the impetus for this growth came from technical innovation. The introduction of turbo-propeller aircraft in the early 1950s, transatlantic jets in 1958, wide-bodied aircraft and high by-pass engines in 1970 and later, advanced avionics were the main innovations. They brought higher speeds, greater size, better unit cost control and, as a result, lower real fares and rates. Combined with increased real incomes and more leisure time, the effect was an explosion in demand for air travel.

Increased demand for air travel led to increased activity for IATA.

Technical work evolved into seven broad areas. Avionics and Telecommunications included the vital air navigation function; Engineering and Environment developed IATA policy on aircraft noise and other emissions; Airports defined airline requirements for airport terminals; Flight Operations worked on safety factors such as minimum aircraft separation standards and air routings; Medical monitored health standards for flight crews and facilitated air travel for disabled passengers; Facilitation attempted to speed the flow of people and goods through airports—particularly, customs and immigration; Security worked on measures to safeguard passengers and cargo by preventing hijacking and sabotage and on minimizing fraud and theft of tickets.

IATA's legal efforts anticipated the effects of new technology associated with the period of rapid growth. It was able to advise the industry on new aircraft and systems, electronic data processing and advances in sales and marketing techniques. Since acts of hijacking and sabotage became more frequent during this period, IATA assisted in the development of the Tokyo, Hague and Montreal Conventions, the first international legal counter-measures.

Automation became common place in airline operations during the 1960s. Standardisation in its use was less common place and IATA began its involvement in perfecting standard message formats for inter-company data exchange, which continues to this day.

The goal has been to save money for the airlines, while enhancing airline service. This philosophy was carried over to many activities during the 1950s and 1960s. Development of the Clearing House was followed by the establishment of Billing and Settlement Plans and Cargo Accounts Settlement Systems—in effect, one-way clearing houses to speed the flow of revenue from agents to airlines. Sales agents were given the opportunity to prove their professional status under an accreditation process and training—in conjunction with the Universal Federation of Travel Agents' Associations and the Federation of Freight Forwarders' Associations—was introduced. The pattern for airline-agent relations was set with the introduction of the Standard Agency Agreement in 1952. There are now nearly 81,000 IATA agents worldwide and 135,000 students have enrolled for IATA-sponsored agency training courses since they were introduced.

International air transport creates special problems of taxation. This was a concern even before World War Two. IATA makes specific challenges to the legality of certain taxes and points out to governments the counter-productive effect of excessive aviation taxation in general. User Charges—payment for using airports and air navigation services—mushroomed during the 1960s and 1970s. IATA's task is to minimise their impact by ensuring that the charges are for facilities actually required, that charges are cost-related and that productivity improvements are built into cost projections. Currency earned by airlines abroad is sometimes blocked by the Central Bank of the country in which it is earned. IATA works to free it, for transfer back to the airline that earned it.

Unit Load Devices, or ULDs—principally, containers—allow rapid, economical cargo handling. IATA Members developed the technical specifications for containers and created a ULD control centre, to keep track of their movements. Until 1955, there was a complete embargo on the air transport of toxic, flammable or corrosive materials. Then IATA developed Dangerous Goods Regulations for their safe carriage. A decade later, Live Animals Regulations provided for suitable standards for the in-flight welfare of animals.

Between the late 1940s and the early 1970s, it could be said that flight was transformed from "a scientific phenomenon to a public utility at the disposal of the entire world." At the same time, the popularisation of civil aviation led to growing demands that the "public utility" view of the industry on the part of governments be modified to encompass greater competition and wider access to markets.

These demands were fuelled by competition for leisure travellers provided by charter companies and the challenge of stimulating demand to fill the new wide-bodied aircraft. Potential demand for air travel had extended well-beyond the original client base of bureaucrats, businessmen and the well-to-do. New travel products and greatly increased price flexibility were needed. As a result, the old basis for conducting IATA Traffic Conferences needed modification.

At the same time the US Government, which had first given IATA Traffic Conferences immunity from national anti-trust law in 1946, began its own review of air transport regulation which, although domestic in scope, was bound to have international repercussions.

The international dimension of the US review resulted in a "Show Cause Order." Hearings took place in 1979, at which IATA was called upon to show cause why anti-trust immunity should not be removed from its Tariff Coordination activities. The outcome mainly affected North Atlantic passenger fares which, for some years, were subject to special rules under an inter-governmental Memorandum of Understanding (MoU) between the US and ECAC states. The present position is that all Tariff Coordination activities continue to be protected.

Eventually, IATA was re-organised on a two-tier basis in October 1979. The tiers comprised:

- Trade Association (technical, legal, financial, traffic services and most agency matters).
- Tariff Coordination (passenger fares, cargo rates, and related conditions and charges).

At present, some 100 members, including the world's largest airlines, continue to participate in Tariff Coordination.

The IATA that emerged since 1979 has put an increasing share of its resources into trade association activities.

IATA also changed the basis of its funding. Much of the association's funding is now done through the marketing of its products and services to Member airlines, other airlines and others in the travel, transport and tourism industry. As a result airlines can access a broader array of professional services, tapping into IATA's expertise. And IATA has greater flexibility in the way it approaches its tasks, while remaining devoted to its non-for profit operations principles.

In many ways, those tasks remained the same as in 1945, or even in 1919. But IATA gave them new relevance and focus by redefining its mission and goals in 1994:

Safety and security: to promote safe, reliable and secure air services.

- *Industry recognition*: To achieve recognition of the importance of air transport worldwide social and economic development.
- *Financial viability*: To assist the industry to achieve adequate levels of profitability, by optimising revenues (yield management) while minimising costs (fuel, charges and taxation).
- *Products and services*: Provide high-quality, value for money, industry required products and services that assist the airlines in meeting the needs of the consumer.
- *Standards and procedures*: To develop const-effective, environmentally-friendly, standards to facilitate the operations of international air transport.
- *Industry support*: To identify and articulate common industry positions and support the resolutions of key industry issues (congestion, infrastructure).

These objectives proved to be relevant and most were carried over to the new millennium, where they still form most of IATA's current mission.

While the 20th century saw the creation and formidable ascension of the air transport industry, the beginning of the 21st century is a time of challenge for the viability of aviation, and a time of major changes.

The terrible events of 9/11 showed the economical fragility of a number of major airlines. The fast increase in jet-fuel prices and additional taxes put an additional burden on airlines.

This made IATA's role as a trade association more crucial as ever. IATA's Director and CEO since 2002, Giovanni Bisignani, completely restructured the association to increase its relevance and speed in driving a broad agenda for industry change.

Since 2002 IATA has spearheaded an industry agenda to enhance safety and improve efficiency.

Two major IATA initiatives were launched, re-shaping important parts of the industry:

- The IATA Operational Safety Audit (IOSA), the first global standard for airline operational safety management.
- The Simplifying the Business programme, using technology to improve customer convenience while reducing costs in five focus areas: E-ticketing (by end 2007, there should be no more paper tickets), Bar Coded Boarding Passes (BCBP), E-freight, Common Use Self-Service (CUSS) and Radio Frequency ID (RFID).

Giovanni Bisignani has also strengthened IATA's position as the voice of the industry, with firm advocacy and lobbying to focus governments on the long-term issues for the viability of aviation including liberalisation, environment and taxation.

IATA's role remains central to ensure that air transport stays one of the most dynamic industries. The market for air transport has changed beyond recognition during the history of the modern IATA. The industry continues to change and IATA is leading it to meet the new needs.

However, one thing has not changed—the necessity to provide service. IATA helps the airlines to achieve this objective as they meet the demands of the 21st century. The airlines have created a modern, interdependent world over the past 60 years. IATA is working to ensure that the world's most exciting industry meets its greatest possible potential.

Cargo Committee (CC)

The Cargo Committee advises the Board of Governors, the Director General and other relevant IATA bodies on all air cargo industry issues, particularly the following:

- Agent/carrier relations,
- Cargo procedures and automation,
- Cargo facilitation, and
- Cargo related regulatory developments.

The Committee meets every six months to consider issues submitted by Member Airlines. Each member receives written notice at least 30 days in advance of the date and an agenda for each meeting. The agenda is limited to issues of significant importance to the airline industry and where there is potential to take action; information only items and activity reporting are kept to a minimum. A summary report of each meeting is sent to all attendees. Decisions are acted upon by the Secretariat. The current Cargo Committee Chairman is Mr. Michael Wisbrun (KLM).

The Cargo Committee comprises 12 members elected from Head of Cargo operations of IATA Member airlines. All members of the Committee are proposed by nomination from the membership at large. The Board of Governors proposes the membership to the Annual General Meeting (AGM). A full list of members and the Cargo Committee mandate are included in this package. Membership of the committee is terminated if a member fails to attend two consecutive meetings without adequate reason.

Heads of Cargo who are not members of the committee are able to attend as observers by prior arrangement.

Mandate

The Cargo Committee shall act as advisor to the Board of Governors (the "Board"), the Director-General, and other relevant IATA bodies (including Conferences) on all air cargo industry policy issues, particularly the following:

- Agency/carrier relations,
- Cargo automation,
- Cargo handling,
- Cargo facilitation, and
- Cargo related regulatory development.

Subject to oversight by and direction from the Board and the Director-General, the Cargo Committee shall identify and assess key policy issues, and develop policies/positions/action plans to resolve such issues; in like manner, it shall establish issue priorities and provide policy guidance to other IATA groups including the

Cargo Conferences. Additionally, the Cargo Committee will promote information sharing and policy campaigning.

In carrying out its functions, the Cargo Committee shall give priority to issues which are cargo related, and which have at least two out of the following three dimensions:

- Aeropolitical,
- Industry cooperation, and
- Industry competition.

In addition, it shall only address issues, which are of significant importance to the airline industry, and where there is a potential to take action.

When addressing IATA Tariff Coordinating conference issues, only members of the Cargo Committee whose airlines are members of the relevant Tariff Coordinating Conferences may participate.

Environment Committee (CNCOM)

The Environment Committee is one of six IATA Industry Committees. ENCOM replaced the former Environment Task Force (ENTAF) on 1 June 2005.

The purpose of ENCOM is to advise the Board of Governors, the Director-General and other relevant IATA bodies on environmental matters, and act as the focal point in IATA on environmental issues. ENCOM is responsible for:

- monitoring, assessing and responding to environmental developments, policies and regulations of concern to IATA Member Airlines,
- developing and recommending common industry positions on environmental issues, and
- advising and implementing strategies to promote IATA positions, amongst regulatory bodies and stakeholders.

Membership is made up of representatives of IATA Member Airlines, who are appointed by IATA's Director-General, with the approval of the Board of Governors.

The following airlines are represented on ENCOM:

- Air France,
- American Airlines,
- Cathay Pacific,
- Delta Air Lines,

- Federal Express,
- Iberia,
- Japan Airlines,
- Lufthansa,
- Qantas,
- Northwest, and
- SAS.

Financial Committee

The Financial Committee is one of IATA's six permanent Standing Committees, reporting to the Board of Governors through the Strategy and Policy Committee. The Committee acts as an advisor to the Board of Governors and the Director-General on all financial matters connected with international air transport.

The Committee delegates much of its ongoing work related to cost control to expert Panels, Task Forces and Working Groups. The Committee also delegates the management of many Financial Services to the IATA Secretariat.

The Committee meets twice each year. Membership consists of 25 representatives elected from among Chief Financial Officers of IATA Member Airlines. Representatives of IATA's Industry Associates may also attend parts of Financial Committee meetings as observers.

Legal Committee

The Legal Committee is one of the IATA Industry Committees established by the Director-General, subject to the approval of the Board of Governors. The role of the Industry Committees is to work closely with IATA management, advising on campaigns, policy issues or questions associated with the operation of industry programmes and to advise the Board on Industry issues, reporting through the Director-General.

The Legal Committee acts as advisor to the Board of Governors, the Director-General and other relevant IATA bodies on legal matters affecting member airlines. The mandate of the Legal Committee, established by the Director-General, subject to the approval of the Board of Governors follows—The Committee:

- acts as an advocate for the air transport industry,
- advises on the legal aspects of IATA Conference issues and services operated by IATA on behalf of the industry,

- provides advice to IATA with respect to outside counsel,
- advises and consents to litigation conducted by IATA on behalf of the industry,
- advises IATA on matters related to the development of international law, and
- takes other action as necessary and appropriate.

Appointments to the Legal Committee take into consideration industry expertise, potential contribution to the Committee's work, regional representation and the views of the Committee members. Appointments are for a renewable periods of up to two years and are subject to approval of the Board of Governors.

Operations Committee (OPC)

Notice to all Users of the Operations Committee Website

Please note that effective July 24th, 2006 the Operations Committee private website was shut down and migrated to a new platform. To access the new site, you must re-register. Once the form is submitted and your profile is approved, you will receive a confirmation e-mail that will contain your new username and password to access the OPC private site.

About the Operations Committee (OPC)

The OPC is one of the six IATA Industry Committees.

It acts as advisor to the Board of Governors and the Director-General on all matters connected with the improvement of safety, security and efficiency of civil air transport.

OPC guides, coordinates and oversees the strategic and tactical work undertaken through working groups and task forces together with the Regional Coordination Groups. The Committee meets twice yearly in spring and fall.

The Committee consists of not more than 20 members appointed by the IATA Director-General on the basis of Member CEO nominations. IATA Strategic Partners may participate in meetings at the invitation of the OPC Chairman.

Chairman for 2006—Tim Steeds (BA)
First Vice-Chairman—Owen Dell (CX)
Second Vice-Chairman—Johnny Woods (SA)
Secretary—Günther Matschnigg (IATA)

Operations Committees Work Groups

(*a*) Flight Operations Group (FOG),
(*b*) Engineering and Maintenance Group, and
(*c*) Technical Fuel Group (TFG).

(a) Flight Operations Group (FOG)

The Flight Operations Group, consisting of 10 representatives from member airlines, governs Flight Operations. In general, the FOG is responsible for developing policies, standards and recommendations for consideration to the IATA Operations Committee (OPC) on all aspects of safe and efficient flight operations.

Flight operations is responsible for keeping good contacts with the member airlines, with industry suppliers and with ICAO. Through these contects developments can be initiated and monitored.

Presently the FOG include 4 Task Forces (TF):

- Aircraft Performance Task Force (APTF).
- SCAP Task Force (SCAPTF).
- Meteorological Task Force (METTF).
- Flight Crew Training Task Force (FCTTF).

The FOG Task Forces comprise representatives from airlines, equipment manufacturers and regulators. They meet once or twice annually to address issues in their respective fields of interest.

Access the Task Forces Private Sites: The private part of the sites is available for IATA members and Strategic Partners.

- In order to access the private sites, you need to register and
- If you are already registered: login to the private site.

Get Performance Input Data When you Need It: The *Airport and Obstacle Database (AODB)* site's centralises airport and runway obstacle data and facilitates the exchange of information related to these issues between airlines, in order to improve operational efficiency and safety. This continuously expanding online library maintains 3000 international, national and regional airport libraries.

FOG Information Exchange (FOG IE): FOG IE enables FOG members to quickly and confidentially exchange information on any matter, from which lessons are learned or for which corrective measures are taken. It is the simplest yet most effective way of improving airline efficiency.

Flight Operations Publications

IATA Technical Operations Policy Manual: The IATA Technical Operations Policy Manual (TOPM) is used to plan and execute a range of technical activities in accordance with IATA policies, positions and recommendations. It serves as a reference and guide for Members, IATA Management and enterprises actively involved in the global development of the airline industry, particularly in deliberations involving State authorities and international organisations.

Flight Simulator Contract Negotiation Guide 2nd Edition: The Flight Simulator Contract Negotiation Guide 2nd Edition (released in November 2000) is intended to facilitate the procurement process for airline operators.

IATA Simulator Documentation Requirements: These requirements for flight simulator documentation are linked to the 'ATA Specification 100—Manufacturer's Technical Data'. They cover requirements for documenting the flight simulator systems (those systems required for simulator operations), and flight simulator cuing systems (simulator systems that provide perceptual cues to the crew in training). This product is sold as a self-contained read/print programme and requires Microsoft Windows.

IATA STD Support Documentation Requirements 3rd Edition: Now extensively revised, this document is intended as a guide for operators, manufacturers, and vendors of aeroplane and FTD equipment to define the scope and content of a support documentation package, its layout, breakdown and delivery form.

Flight Simulator Evaluator Pilot Guide 1st Edition: The Flight Simulator Evaluator Pilot Guide 1st Edition (released in January, 2001) is now available. This publication describes training and qualification requirements for flight simulator evaluator pilots. It outlines recommended syllabii for a range of courses which address the development, acceptance, and qualification of most synthetic training devices.

(b) Engineering Maintenance Group (EMG)

The EMG develops policies, standards and recommendations for consideration by the *IATA Operations Committee* in all aspects of safe and efficient aircraft engineering and maintenance activities, airworthiness and reliability issues, aircraft recovery, aircraft performance and avionics. Additionally, it develops and promotes related IATA positions to ICAO, regulatory, industry and other relevant organisations.

EMG Information Exchange: The Engineering and Maintenance Group Information Exchange (EMG IE) allows members to quickly and confidentially exchange information on any matter with technical implications, from which lessons have been learned or for which corrective measures taken.

This exchange is widely recognised as being one of the simplest and yet most effective activities for improving airline efficiency.

(c) Technical Fuel Group (TFG)

The objective of the Technical Fuel Group (TFG), former Aviation Fuel Working Group, is to ensure appropriate industry action on all safety, environmental and technical aspects of aviation fuel. The TFG works closely with the *Commercial Fuel Group (CFG)* on many issues, including the holding of the bi-annual Aviation Fuel Forum.

Membership

- The TFG is composed of ten (10) member airlines.
- It includes a supplier representative, elected by suppliers in the Strategic Partnership programme.
- A Chair and one Vice-Chair is elected by the ten member airlines to serve a period of two years.
- Each member has one vote.

Terms of Reference

- Work with industry suppliers and service providers to ensure 'on specifications' fuel is supplied at all airports.
- Investigate all fuel related incidents and accidents and make recommendations to enhance the safety of the fuelling operation.
- Develop and publish international standards, procedures and guidelines for Jet fuel.
- Work with international and national standard setting bodies to pursue harmonisation efforts in view of achieving one global standard for Jet fuel specifications, its safe handling and quality control.

- Encourage the participation of airlines in the IATA Fuel Quality Pool to achieve economic benefits, meet safety and technical objectives and comply with regulatory requirements.
- Support all industry effort to achieve IATA fuel efficiency and emissions goals.
- Investigate the potential offered by alternative fuels.
- Ensure effective working relationships are developed and maintained with other groups, both within and external to IATA. Of particular importance are the relationships with the supplier community, and the Commercial Fuel Group.
- Develop work programmes and report to IATA Management on achievements.
- Plan and organise the Aviation Fuel Forum together with the Commercial Fuel Group to ensure full airline participation.

Industry Affairs Committee (IAC)

The Industry Affairs Committee advises the Board of Governors, the Director-General and other relevant IATA bodies on the regulatory aspects of commercial/industry affairs and aeropolitical matters affecting international passenger air transport.

More specifically, it assists IATA in identifying and developing industry policies and positions, supervising policy implementation and promoting policy campaigning. The IAC has specific responsibility for ensuring that IATA's multi-lateral interline system does not contravene competition laws and for ensuring that proposed consumer protection regulation does not result in adverse financial or practical implications for the air transport industry.

How it Works

The Committee meets twice each year, usually in the spring and the autumn. Agendas are limited to issues of significant importance to the airline industry and where there is potential to take action; information only items and activity reporting are kept to a minimum. Decisions are acted upon by the Secretariat with support from the IAC members.

IAC Members

The IAC comprises 20 members appointed by the IATA Director-General on the basis of Member CEO nominations. Membership of the committee is terminated if a member fails to attend two consecutive meetings without adequate reason. Representatives of IATA Member airlines who are not members of the committee are able to attend as observers by prior arrangement.

Private Site

The private site provides a forum for information exchange and discussion for IAC members and for all representatives of IATA members that work in the regulatory/industry affairs field.

RULES AND REGULATIONS OF THE INDUSTRY COMMITTEES

I. Role and Mandate

(1) Industry Committees shall be established by the Director-General, subject to the approval of the Board of Governors. Their role shall be to:

  (*i*) Work closely with IATA management, advising on campaigns, policy issues or questions associated with the operation of industry programmes.

  (*ii*) Advise the board on industry issues, reporting through the Director-General.

(2) The mandates of the Individual Industry Committee will be established by the Director-General, subject to the approval of the Board of Governors. The mandates are detailed in Annex 1.

(3) Policy recommendations developed by the Industry Committees shall be presented to the board following adequate industry consultation.

II. Membership

(1) Industry Committees shall be comprised of six to 20 members. The number of observers is limited to six. There are no alternates.

(2) Members of the Industry Committees shall be appointed by the Director-General on the basis of Member

nominations. Nominees shall have a major responsibility for the strategic formulation and execution of airline policy.

(3) The appointments shall take into consideration industry expertise, potential contribution to the Committee's work, regional representation and the views of the Industry Committee concerned.

(4) Appointments shall be for a renewable period of up to two years and are subject to approval of the Board of Governors.

(5) Each member shall act as a representative of the Association, and not as a representative of the airline Member.

(6) The Committee Secretary will be appointed by the Director-General.

(7) Each Committee shall elect its Chair and Vice-Chair, whose term shall be for two years.

(8) Each Committee may establish a Steering Committee to coordinate required activity between meetings with IATA management.

(9) Membership shall terminate if the member misses two consecutive meetings, or if the member leaves his position in the airline. The Director-General may fill vacancies on an interim basis.

III. Meetings and Procedures

(1) Industry Committees shall meet as required, including by teleconference, video conference, or other electronic means.

(2) Meetings shall be called by the Secretary, in consultation with the Committee Chair.

(3) The Secretary shall normally give thirty days' notice of the meeting; at least ten days notice shall be given for a special meeting.

(4) A majority of the Committee members shall constitute a quorum at any meeting.

(5) The agenda of any meeting shall include any matter proposed by:

(*i*) the Director-General;

(*ii*) any member of the Committee; or

(*iii*) any Member of the Association—the Member proposing an item shall be entitled to attend the meeting of the Industry Committee; and

(*iv*) The Committee Secretary.

(6) The Committees shall work on the basis of consensus. Should a vote be required, each Committee member shall have one vote, and decisions shall be taken only by a majority vote of the members present and voting. Issues may be considered and decisions taken via voice votes, via email or via any means agreed upon by the Committee.

(7) Working Groups and Ad Hoc Task Forces may be established by Industry Committees, subject to the approval of the Director-General. The Working Groups shall:

(*i*) Provide technical advice on an on-going basis.

(*ii*) Work with IATA management on campaigns and issues.

(8) Working Groups shall be limited to a maximum of 10 members who shall be appointed by the relevant Industry Committee or its Steering Group, as applicable. The Committee Secretary shall provide the coordination between the activity of the Working Group and the Industry Committee.

(9) Ad Hoc Task Forces may be established to address specific projects and shall be disbanded once the project is completed.

(10) Each Committee may establish its working procedures consistent with these Rules.

(11) These Rules and Regulations may from time to time be amended by the Board.

ANNEX 1
MANDATES IATA INDUSTRY COMMITTEES

Cargo Committee

The Cargo Committee shall act as advisor to the Board of Governors, the Director-General, and other relevant IATA bodies on all air cargo industry policy issues and develop policies/ positions/action plans to resolve such issues. Areas of activity include:

(*i*) Agent/carrier relations.
(*ii*) Cargo automation.
(*iii*) Cargo handling.
(*iv*) Cargo facilitation.
(*v*) Cargo-related regulatory development.

Financial Committee

The Financial Committee shall act as advisor to the Board of Governors, the Director-General, and other relevant IATA bodies on all financial matters connected with international air transport and in particular the following:

(*i*) Revenue Accounting.
(*ii*) Taxation.
(*iii*) User Charges.
(*iv*) Fuel Trade.
(*v*) Clearing House.
(*vi*) Insurance.
(*vii*) Industry Accounting Guidelines.
(*viii*) Statistics.
(*ix*) Currency.
(*x*) Airline Cost Issues.

Industry Affairs Committee

The Industry Affairs Committee shall act as advisor to the Board of Governors, the Director-General, and other relevant IATA bodies on all industry affairs and aeropolitical matters connected with international passenger air transport. The Industry Affairs Committee shall develop industry positions, supervise policy implementation, and promote campaigning, particularly in the following areas:

(*i*) Customer service, including passenger and airport services.
(*ii*) Facilitation.
(*iii*) Governmental, inter-governmental and other air transport policy.
(*iv*) Marketing and distribution.
(*v*) Scheduling.
(*vi*) Tariffs and pricing.
(*vii*) Anti-trust immunity and competition law exemptions.

Operations Committee

The Operations Committee shall act as advisor to the Board of Governors, the Director-General, and other relevant IATA bodies on all matters that relate to the improvement of safety, security and efficiency of civil air transport. This will include, but not be limited to matters that relate to:

(*i*) Airline safety.
(*ii*) Flight operations and global air traffic management.
(*iii*) Engineering and maintenance.
(*iv*) Security.
(*v*) Aviation infrastructure.

Environment Committee

The Environment Committee shall act as advisor to the Board of Governors, the Director-General, and other relevant IATA bodies on environmental matters and act as the focal point in IATA on environmental issues. Specifically, the Environment Committee shall:

(*i*) Monitor, assess and respond to environmental developments, policies and regulations of concern to IATA Member airlines.
(*ii*) Develop and recommend common industry positions on environmental issues.
(*iii*) Advise and, as necessary, implement strategies to promote IATA positions with regulatory bodies and stakeholders.

Legal Committee

The Legal Committee shall act as advisor to the Board of Governors, the Director-General, and other relevant IATA bodies on legal matters affecting Member airlines. The Committee shall:

(*i*) act as an advocate for the air transport industry,
(*ii*) advise on the legal aspects of IATA Conference issues and services operated by IATA on behalf of the industry,
(*iii*) provide advice to IATA with respect to outside counsel,
(*iv*) advise and consent to litigation conducted by IATA on behalf of the industry,
(*v*) advise IATA on matters related to the development of international law, and

(*vi*) take other action which is considered necessary and appropriate.

IOSA Oversight Committee (IOC)

The IOSA Oversight Committee (IOC) is an entity within the IATA governance structure set-up to ensures that:

- The IOSA Programme Team uses processes compliant with recognized quality standards;
- The IATA member airlines play an active role in the IOSA programme; and
- The IOC roster is representative of the global industry.

The IOC membership is comprised of representatives from twenty-five Member Airlines, and ten Regulatory Authorities. The IOC reports to the Operations Committee, which in turn, reports to the Board of Governors of IATA.

The next meeting of the IOSA Oversight Committee (IOC/9) will be held in Paris, France on September 12th and 13th, 2007.

IOSA Regulatory Authorities

This site is restricted for regulatory authorities togather information on the IOSA programme.

The IOSA programme encourages regulatory authorities to:

- Participate in the growth of IOSA through the IOSA Oversight Committee.
- Accept IOSA as the internationally recognized operational audit programme.
- Encourage Airlines to seek registration.

If you are with a regulatory authority involved with IOSA and do not have a password, click on the Register button. The next meeting of the IOSA Oversight Committee (IOC/9) will be held in Paris, France on September 12th and 13th, 2007.

IATA Operational Safety Audit (IOSA)

The IATA Operational Safety Audit (IOSA) programme is an internationally recognised and accepted evaluation system designed to assess the operational management and control systems of an airline. IOSA uses internationally recognised quality audit principles, and is designed so that audits are conducted in a standardised and consistent manner.

Inherent in the IOSA programme is a degree of quality, integrity and security such that mutually interested airlines and regulators can all comfortably accept IOSA audit reports. As a result, the industry will be in a position to achieve the benefits of cost-efficiency through a significant reduction in audit redundancy.

With the implementation and international acceptance of IOSA, airlines and regulators will achieve the following benefits:

- The establishment of the first internationally recognised operational audit standards.
- A reduction of costs and audit resource requirements for airlines and regulators.
- Continuous updating of standards to reflect regulatory revisions and the evolution of best practices within the industry.
- A quality audit programme under the continuing stewardship of IATA.
- Accredited audit organisations with formally trained and qualified auditors.
- Accredited training organisations with structured auditor training courses.
- A structured audit methodology, including standardised checklists.
- Elimination of audit redundancy through mutual acceptance of audit reports.
- Development of auditor training courses for the airline industry.

IOSA Audit Organisations (IAO) Private Site

The IOSA Audit Organisations private site allows the sharing of information on IOSA programme among the organisations formally accredited by IATA to conduct IOSA audits.

The next meeting of the IOSA Oversight Committee (IOC/9) will be held in Paris, France on September 12th and 13th, 2007, with the AO meeting being held on September 10th and 11th.

Passenger XML Working Group (XMLWG)

The Passenger XML Working Group was established by the IATA Passenger Services Conference with the following Terms of Reference:

- To identify and develop business requirements for airline internet data interchange applications based on current airline practices and procedures;
- To develop airline internet data interchange message specifications to meet the defined business requirements;
- To develop implementation guidelines for the airline internet message specifications;
- To propose/submit additions, changes and deletions to the IATA airline internet data dictionary;
- To liaise—where appropriate—within IATA and with other internet data standards development organisations;
- To submit the data interchange standards to ECDG for endorsement;
- To submit the data interchange standards to PADIS Board for endorsement; and
- To provide work plans and status/progress reports to PADIS Board.

Fuel Data Standards Work Group

Adoption of industry data standards makes the application of technology more affordable, easier and faster to deploy, more accurate and more effective.

This is particularly relevant in the area of electronic invoicing and provision of supporting details. Invoice standards deliver substantial benefits to both the Buyer and Supplier of services. These include:

- Improved data quality and accuracy.
- Improved staff productivity.
- Reduced number of disputes and queries.
- Improved timely invoice settlement and cash management.
- Improved customer service.
- Reduced cost of AP/AR function.
- Reduced numbers of IT adaptors needed for the exchange of e-data with business partners.

The Aviation Fuel Data Standards Group develops and maintains industry data standards for the efficient and effective data exchange across the aviation fuel industry.

In addition to standards the Group maintains commonly

associated terms, definitions and codes as required to fulfill message requirements.

Examples of the Group's work include the XML Fuel Invoice Standard and the XML Fuel Transaction Standard. These standards are available to interested parties upon request.

The Group comprises Airline, Fuel Supplier and Into-Plane Service Provider representatives in charge of maintaining the Fuel Data Standards. You can find here the group's. Other interested parties may also participate in the Group's activities upon request on a case by case basis.

IATA InvoiceWorks

Welcome to IATA InvoiceWorks. Welcome to the world of electronic invoicing.

Established by the International Air Transport Association (IATA) and powered by BearingPoint, IATA InvoiceWorks provides an electronic invoicing service for the air transport industry to cut costs, time and effort for sending, receiving and processing invoices.

IATA InvoiceWorks is the world's only Internet-based electronic invoicing service focused on the needs and demands of the global air transport industry. Designed by and for aviation experts, IATA InvoiceWorks provides a unique combination of global reach and functionality richness that no alternative can match.

Over 260 airlines and more than 5,550 industry suppliers are already linked and enjoying the benefits of electronic invoicing through the advantages of IATA InvoiceWorks.

Leading companies understand that invoicing is a strategic process. A well-managed invoicing process improves financial performance, enforces internal controls and streamlines the order-to-pay cycle. Electronic invoicing provides dramatic, ongoing benefits that can cut related costs up to 75 per cent. Whether delivering or receiving invoices, a range of proven benefits are achieved, including:

- Optimized working capital management.
- Increased capture of supplier discounts.
- Fewer payment delays.
- Strengthened audit and regulatory compliance.
- Reduced processing costs.

- Reduced labour costs and increased staff productivity.
- Improved customer service.
- Strengthened important trading relationships.

Air Navigation Services Invoice Task Force (ANSITF)

Use of agreed industry standards for the exchange of data between business partners can drive efficiency and cost reduction for all stakeholders.

This is particularly relevant in the area of electronic invoicing and provision of supporting details. Invoice standards deliver substantial benefits to both the Buyer and Supplier of services. These include:

- Improved data quality and accuracy.
- Improved staff productivity.
- Reduced number of disputes and queries.
- Improved timely invoice settlement and cash management.
- Improved customer service.
- Reduced cost of AP/AR function.
- Reduce numbers of IT adaptors needed for the exchange of e-data with business partners.

The Air Navigation Services Invoice Standard and Data Dictionary facilitate the exchange of billing data between Air Navigation Services providers and Airlines and simplify the adoption of electronic invoicing in the air transport industry.

ANSITF comprises of Airline and Air Navigation Service provider representatives is in charge of maintaining the Air Naviation Services Invoice Standard and Data Dictionary. If your company is interested in joining the ANSITF.

Adopting the Air Navigation Services Invoice Standards

The Air Navigation Services Invoice Standard and Data Dictionary will be released soon.

Airport Invoice Standards Task Force (AITF)

Use of agreed industry standards for the exchange of data between business partners can drive efficiency and cost reduction for all stakeholders.

This is particularly relevant in the area of electronic invoicing and provision of supporting details. Invoice standards deliver substantial benefits to both the Buyer and Supplier of services. These include:

- Improved data quality and accuracy.
- Improved staff productivity.
- Reduced number of disputes and queries.
- Improved timely invoice settlement and cash management.
- Improved customer service.
- Reduced cost of AP/AR function.
- Reduced number of IT adaptors needed for the exchange of e-data with business partners.

The Airport Invoice Standard and Data Dictionary facilitate the exchange of billing data between Airport services providers and Airlines and simplify the adoption of electronic invoicing in the air transport industry.

Airport Invoice Task Force (AITF)

AITF comprises of Airline and Airport service provider representatives is in charge of maintaining the Airport Invoice Standard and Data Dictionary. If your company is interested in joining the AITF, please contact the E-Invoicing Services team at *Invoicestandards@iata.org.*

Adopting the Airport Invoice Standards

The Airport Invoice Standard and *Data Dictionary* have been published and are now available for download by any Airport services providers and Airlines that elect to adopt it.

Index